mushroom

mushroom

johnny acton & nick sandler

with photography by jonathan lovekin

TED SMART

To Jack

This edition produced for The Book People Ltd
Hall Wood Avenue, Haydock, St Helens WA11 9UL

First published in Great Britain in 2001
by Kyle Cathie Ltd,
122 Arlington Road, London NW1 7HP
general.enquiries@kyle-cathie.com
www.kylecathie.com

ISBN 1 85626 423 8

A CIP catalogue record for this title is available from the British Library.

Designer: Geoff Hayes
Index: Helen Snaith
Production: Lorraine Baird & Sha Huxtable
Repro: Colourscan
Printed and bound in Singapore by Tien Wah Press

There are many people without whose assistance we couldn't even have begun to do this book. In roughly alphabetical order, then, and at the certain risk of leaving out somebody important, we would like to thank the following: Kyle Cathie and her staff for taking on the project in the first place, Penny David for her highly proactive editing, Hugh Fearnley-Whittingstall for several useful leads, Conky Galitzine for information on Reishi mushrooms, Geoff Hayes for his inspired design, Peter Jordan for the benefit of his mycological wisdom, Livesey Brothers Mushroom Farm in Leicestershire for an invaluable insight into commercial fungus production, Dick Peebles for leading us to all sorts of species in Scotland, Roger Phillips for his very generous quote, and Mike and Peter de Stroumillo of Wild Harvest for supplies, tips, contacts and coffee, and their quote.

contents

introduction

Mushrooms have always had something magical about them, from the seemingly miraculous ways in which they grow through their medicinal properties to their sometimes transcendentally exquisite taste.

Because of the mysterious, apparently capricious way in which they pop up, mushrooms are an excellent antidote to the controlled, predictable world in which most of us now live. Much about them remains thrillingly outside the grasp of science. There is something wild and pagan about fungi that makes them fascinating enough on their own. But they can also be wonderfully good to eat.

Mushroom is written by two men who are equally enchanted by the kingdom of fungi, but who come at the subject from very different starting points. Nick looks at mushrooms with the cool eye of the experienced chef and forager. He is, in other words, always thinking about his stomach (or at least about everyone else's). Johnny, on the other hand, is more the wide-eyed innocent when it comes to looking for mushrooms systematically, let alone deciding what to do with them next. His particular interest is in their cultural significance, their role in history and in myth. He also happens to think they look rather nice.

What first brought us together, though, was our shared passion for soup. The outcome, which we hope has made some tiny imprint on the consciousness of the nation, was Soup Works, a chain of funky West End restaurants intent on overthrowing the tyranny of the lunchtime sandwich. We also wrote a recipe book, *Soup*. Noting how much fun we had in the process, we thought it would be a good idea to collaborate again. The subject matter chose itself. We had discovered our mutual love of mushrooms in the early days, while arguing about what to put in the soups.

At the heart of *Mushroom* are more than seventy recipes, ranging from inspired creations to mushroom classics foraged from all over the world. The most popular fungi are covered in all the depth they deserve. But this is also a book that relishes some of the less familiar varieties; partly to reflect people's growing culinary adventurousness, partly because of the increase in exotic species available in the shops, but mostly just for fun.

Mushroom, however, sets out to be more than just a fantastic and adventurous cookbook. It is also a celebration of mushrooms in the wider sense.

This book not only examines the fascinating impact fungi have had on every aspect of human culture from the murder of emperors to the true origin of the myth of Santa Claus. It also seeks to transmit the thrill of the chase. In common with fishermen everywhere, true fungophiles have a treasure-trove of stories about their obsession, and jealously guard the secrets of their favourite foraging sites. *Mushroom* is packed with choice examples, like an ancient ring of St George's Mushrooms in Salisbury Plain which measures over two miles across.

Each chapter of the book takes one aspect of the wonderful world of edible mushrooms and explores it in depth. The multi-angled approach incorporates material from worlds as diverse as those of history, economics, folklore, anthropology and myth. The result, we hope, is that the recipes at the core of the book are not disconnected, like so many others, from the life of their raw materials. These are recipes imbued with the spirits of both the mushrooms themselves and of the people who love and use them. And to do justice to the beauty of the subject, we enlisted a mighty fine photographer to do the shots. No one who has stood in the rain watching Jonathan Lovekin crouching over a Puffball for several hours could doubt that he was born to do the job.

Many mushroom books have approached the subject purely in terms of their culinary uses or from the worthy but dry perspective of taxonomy (identification and classification). *Mushroom* aims to be different. This is about mushrooms 'in the round'.

'... mushrooms are an excellent antidote to the controlled, predictable world in which most of us now live.'

chapter 1

mushrooms of the gods

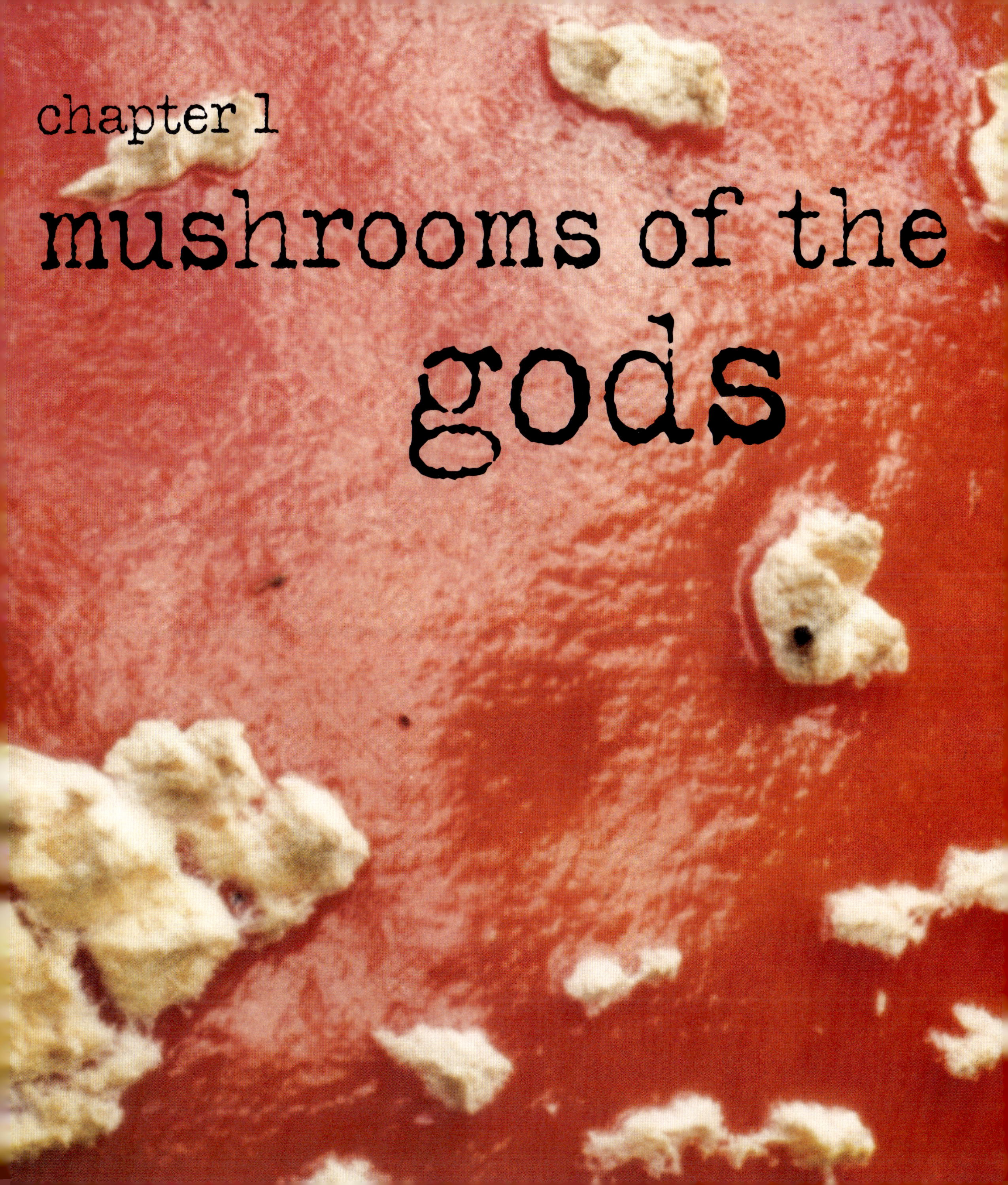

For much of human history, the origin of mushrooms was a mystery, in the original, quasi-religious sense of the word. As far as plants and animals were concerned, the mechanics of reproduction were relatively obvious – had they not been, agriculture would have been a non-starter. But the various forms of fungi seemed, even on close inspection, to come from nowhere. Little wonder that they were often seen as direct manifestations of the spirit world.

When we consider that even in the scientifically sophisticated world of today much about the growth and propagation of mushrooms is not understood, it is easy to appreciate the impression their appearance made on our forebears. In the absence of plausible 'naturalistic' explanations, pre-scientific cultures attributed the genesis of mushrooms to the supernatural.

One surprisingly widespread theory, with adherents as far afield as Mexico, India and Ancient Greece, held that mushrooms were the result of lightning bolts (themselves of divine origin) striking the ground. This belief gives us some idea of how awesome they must have seemed to our ancestors. Mushrooms were dangerous, not quite terrestrial, the embodiments of terrifying natural powers.

Among the Germanic peoples, the distinctive red and white Fly Agaric (*Amanita muscaria*) was believed to grow from blood-flecked foam from the mouths of horses belonging to the god Wotan (Odin) and his companions, who charged through the forest once a year to escape a pack of demons. People everywhere noted the mysterious circles in which some species grew, and put them down to the footsteps of rings of dancing fairies. Anyone who dared step inside these rings risked being taken by the little people.

Even now, wild mushrooms retain an aura of the supernatural. Neither plant nor animal, their otherworldliness is often enhanced by bewitching shapes and colours. And like apples in the world of myths and fairy-tales, mushrooms are fascinatingly ambiguous. They beguile with the promise of sensual delight, but to the unwary they can be deadly. Little wonder that our ancestors' attitude to fungi ranged from reverence to terror.

Logically, we would expect the first things to have struck early man about mushrooms to have been that some were good to eat – the subject of most of what follows – and some downright poisonous. All cultures presumably had to go through an initial 'Russian Roulette' phase with the local fungi, and many believe the outcome determined whether their descendants became generally mycophile (like the Poles and Italians) or mycophobic (like the British). But in fact, it was two other properties that seem to have left the deepest mark on the primitive mind: the power of some varieties of mushroom to induce trances and visions, and the startling resemblance of others to the human penis.

There is abundant evidence that fertility and the means of securing it were major obsessions for early societies. These range from the lingams (sacred carved phalluses) of Ancient India to the huge chalk Cerne Abbas giant in Wiltshire, a figure who is quite definitely pleased to see us. In this context, it is easy to see how the more phallic mushrooms would have seemed to incarnate the very force that made the world go round. As a result, it was inevitable that they would be revered and treasured. They were also, on the principle of 'like for like', highly sought after as aphrodisiacs.

Instructively, even in the modern world, the most highly valued mushrooms of all are the most phallic examples of the Matsutake or Pine Mushroom (*Tricholoma matsutake*). Pristine near-identical pairs can fetch $10,000 or more in Japan, where they are presented as high-status corporate gifts or wedding presents. They are never, however, given to women. Indeed, in the eleventh century the women of the Imperial Court in Kyoto were forbidden from even pronouncing the Matsutake's name.

So revered are Matsutake in Japan that people will willingly pay for the privilege of picking them, even if they are obliged to hand them in to the landowner before they depart. The demand created by this love affair can sometimes have fatal consequences. In the woodlands of Oregon, competition for the local version of the Matsutake has led to several murders among ethnic Laotians looking to tap the lucrative export market.

Matsutake Stir-fry

Serves 4

What is one to do with what is arguably the most expensive form of fungus on earth, should one be lucky enough to have some fall into one's hands? This superb Japanese stir-fry is a possible answer to one's enviable dilemma.

150 ml/¼ pint chicken or vegetable stock
100g/4oz Matsutake, roughly chopped
2 tbs vegetable/sesame oil mix
50g/2oz sliced spring onion
100g/4oz shredded carrot
100g/4oz baby corn, roughly chopped
1 tsp finely chopped ginger
1 clove garlic, finely chopped
2 tbs soy sauce
A splash of lime juice
1 tbs mirin or sweet sherry
1 tbs brown miso paste
150g/5oz water spinach or regular spinach, roughly chopped, including stalks

Warm up the stock and soak the mushrooms in it for 1 hour so that they absorb its flavour. Drain the mushrooms and reserve the stock.

Pour the oil into a wok or large saucepan, turn the heat right up, and fry off the spring onions, carrot, baby corn, ginger, garlic and Matsutake for about 5 minutes, tossing frequently.

Combine the stock, soy sauce, lime juice, mirin and miso paste and pour into the wok. Bring to the boil and reduce a little. Add the spinach and cook for a couple of minutes.

Serve on a bed of rice.

'... the Fly Agaric has had a rather remarkable influence on both history and religion.'

Turning now to the ancient appetite for hallucinogenic mushrooms, we can gain an interesting insight from the habits of the Kuma tribe of the Highlands of New Guinea, a people who until recently were effectively living in the Stone Age. The Kuma are known to be inordinately fond of four types of fungus which grow on rotting logs, collectively known as 'nonda'. One anthropologist lists the effects of nonda as including 'double vision, cold shivers, loss of speech, intense hallucinations, and most importantly a tremendous frenzy that may lead to casualties'. Exactly why these symptoms should appeal to the Kuma is left unclear, but our anthropologist is in no doubt in general terms. 'The use of hallucinogens is mainly for entertainment for the Highlanders,' he concludes with a straight face.

Father William Ross, a missionary writing in 1936, may have been nearer the mark when he described nonda as being 'used before going out to kill another native, or in times of great excitement, anger or sorrow'. The Kuma are not the only ones to have eaten wild mushrooms in order to reach catharsis, or to enhance their performance in battle. The Berserkers, a ferocious caste of bearskin-wearing Viking royal bodyguards, are believed to have used Fly Agaric (see below) for the same purposes. One of the effects of this fungus is the almost complete eradication of fear. This would tie in well with the Berserkers' terrifying reputation in battle.

If psychedelic mushrooms were the lager of the ancient world to some cultures, to others they were sacred paths to the realm of the gods and spirits. Perhaps the most historically important of all the hallucinogenic mushrooms is the Fly Agaric (*Amanita muscaria*), the archetypal red and white toadstool that supports the posterior of a million garden gnomes.

Some academics have got quite carried away with the Fly Agaric. Particularly John Allegro, a former Lecturer in Theology at Manchester University, who decided that the Bible was basically a coded collection of mushroom myths, and its god a giant mushroom in the sky. There are those who suspect that Allegro had indulged rather too heavily in his pet subject himself. But there is, in fact, no denying that the Fly Agaric has had a rather remarkable influence on both history and religion.

Many scholars, for example, believe that the 'Divine Soma', the god-narcotic worshipped and consumed in religious ceremonies by the original Aryan conquerors of India, was actually the Fly Agaric. Excavations in what is now Turkmenistan have revealed ancient shrines from the second millennium BC containing huge ceramic bowls. Perhaps these were originally Soma dispensaries.

Something similar seems to have gone on at Eleusis in Ancient Greece. Here, a couple of hours' walk outside Athens, thousands of pilgrims gathered every autumn to be initiated into the sacred Eleusinian Mysteries. No one knows for certain what happened there: those who participated were sworn to secrecy on pain of death. What is known is that the ceremony was open to anyone who spoke Greek, including women and slaves, and that the initiates congregated in a great hall called the Telesterion (the Great Hall of Mysteries). It was here that they imbibed the Kykeon, a life-changing potion served from a hidden chamber, which may or may not have contained Fly Agaric. Plato, Aristotle and Sophocles are among those known to have taken part.

Closer to modern times, even Allegro's sternest critics would concede that the Tree of Knowledge depicted in a thirteenth-century fresco in the church at Plaincouralt in France is a branched *Amanita muscaria*. Still more significant, at least for the under-eights, are the extraordinary consequences of a predilection for the Fly Agaric among certain reindeer-herding tribes of Lapland and northern Siberia. Making use of the fact that the psychoactive ingredient in the mushrooms can be passed on through the urine, for certain rituals the shamans of these tribes dress up in the red and white livery of their sacred food and distribute samples of *Amanita* to their tribesmen via their bladders. Incredibly, this seems to be the real origin of the story of Father Christmas.

While we would strongly advise against experimenting with the Fly Agaric, all the hoohah surrounding it does make one a little curious. So what does it actually do to those who consume it? About one and a half hours after ingestion, we are reliably told, the experience commences with involuntary twitching and shaking, combined with a high level of euphoria. In larger doses, this can tip over into apparent raving madness. This excitable phase is followed by a period dominated by vivid hallucinations, whereupon the taker falls into a deep sleep. For one such experience, the Koryak people of eastern Siberia were said in the 1930s to be prepared to part with one reindeer.

Other fungi that have played an important role in religious ritual are those containing the potent alkaloid psilocybin, among them the Liberty Cap (*Psilocybe semilanceata*) beloved of hippy students. Psilocybin mushrooms have been sacred in parts of Central and South America since at least 1000 BC, to go by the evidence of the oldest known 'mushroom stones' from those parts.

The Conquistadors of the sixteenth century were well aware of the ritual use of mushrooms among the Mexican Aztecs. The missionary Bernardino de Sahagún has left us the following contemporary description:

'The first thing to be eaten at the feast were small black mushrooms that they call *teonanacatl*, which bring on drunkenness, hallucinations and even lechery. They ate these before dawn . . . with honey, and when they began to feel the effects, they began to dance. Some sang and others wept . . . When the drunkenness of the mushrooms had passed, they spoke with one another of the visions they had seen.'

Recently a carpet of the most potent of magic mushrooms appeared on a racetrack in the South of England – the Wavy-Capped Magic Mushroom. It was estimated to consist of hundreds of thousands of fruit bodies of *Psilocybe cyanescens*, the mycelium imported in wood chips from North west America.

Teonanacatl means nothing less than 'flesh of the gods'. The mushroom it refers to is probably either *Psilocybe caerulescens*, or *Psilocybe mexicana*. Both are native to Central America and highly psychoactive.

Lest we in Europe start to feel that the religious use of mushrooms is somehow 'primitive', we should bear in mind that it may have played an important role in our own national histories. It has often been suggested that the traditional British suspicion of wild fungi derives from a time when mushrooms were forbidden to all but the druids, to prevent psychoactive varieties from falling into the wrong hands.

Magic Mushroom Omelette with Jalapeño Chilli

Serves 1

This recipe is included purely for curiosity, you understand. It is illegal in this country to prepare psilocybin mushrooms in any way, and we could not possibly condone anyone actually making this dish, let alone eating it. It is rather good, though, even without the miscreant mushrooms, whose taste it is designed to disguise.

A friend of ours used to preserve magic mushrooms in marmalade. Another relatively painless way to take them is said to be in tea.

1 level tbs butter
2 green jalapeño chillies (seeds removed), finely chopped
1 clove garlic, finely chopped
20 imaginary magic mushrooms, finely chopped
2 eggs, broken into a small bowl, a little salt and pepper added, then whisked
1 tsp chopped coriander

Melt the butter in a small, nonstick omelette pan. Fry the jalapeño chillies and the garlic on a moderate heat for a minute.

In your imagination, add the mushrooms and continue to cook for a few moments.

Turn the heat up and add the eggs. As they cook, whisk vigorously with a fork. They will cook in 1 minute. Turn on to a plate, folding over as you do so.

Garnish with a little coriander.

Before leaving the subject of heavenly fungus for good, mention should be made of *Lecanora esculenta*, a desert lichen with the singular habit of uprooting itself and blowing about in the wind. Many scholars believe *Lecanora esculenta* is identical to Manna, the miracle food collected by the wandering Israelites in the book of Exodus. It certainly grows in relative abundance in the Middle East, where people have long been known to search for it in times of famine. In parts of Northern Iran and Turkey it is even sometimes mixed with meal to make bread. The relevance of all this becomes clear when we consider what lichen actually is: a happy symbiosis of algae and fungus . . .

One rung below divine mushrooms were those reserved for royalty. We know that in Ancient Egypt the pharaohs issued decrees forbidding commoners to even *possess* mushrooms, so scarce were they in the prevailing hot, dry conditions. Meanwhile, in Japan prior to the eighteenth century consumption of the Matsutake was strictly limited to the Imperial Court. But perhaps the most celebrated of the Royal Mushrooms is a delicious orange-capped species which we unfortunates in Northern Europe can only find when we journey southwards to the shores of the Mediterranean. Given global warming, it may be only a matter of time before the Caesar's Mushroom moves up north.

Caesar's Mushroom (*Amanita caesarea*) was, as its name suggests, particularly popular with the Roman emperors. Somewhat confusingly, in view of what we know as the quite separate *Boletus* family, they were known as boleti and cooked in a special vessel called a *boletaria*. So prized were they that the first-century poet Martial was moved to write: 'Gold and silver dresses may be trusted to a messenger, but not *boleti*.'

As well as possessing a delicious sweet, nutty taste, Caesar's Mushrooms had the added cachet of being closely related to some of the most deadly fungi of all, including the Death Cap (*Amanita ph*danger associated with them only enhanced their glamour. It also spelled the end for the Emperor Claudius, who was murdered by his wife Agrippina in AD 54 with a dish of the imperial mushrooms soaked in the juice of Death Caps. A salutary warning with which to stop this chapter. Take great care not to confuse Caesar's Mushrooms with elderly Fly Agarics *alloides*). As with the potentially lethal *fugu* (puffer fish) eaten in Japan, the element of whose flecks have been washed off by the rain, or your dinner party may turn out to be more than you bargained for.

Caesar's Mushroom and Aubergine Salad

Serves 4

Last year, we found vast numbers of Caesar's Mushrooms in the South of France, poking out of the bases of young sweet chestnut trees. We then moved on to a treeless hill, where they were growing in equal abundance. We wound up with so many that we resorted to freezing some of them whole. This turned out to work quite well, provided we cooked them straight from the freezer – otherwise they tended to become mushy during cooking.

The salad
8 baby aubergines
Olive oil for brushing
Salt and black pepper
A small handful of pine nuts
8 medium Caesar's Mushrooms with open caps
A couple of sprigs of thyme, leaves separated from stalks
2 handfuls of rocket leaves
2 handfuls of baby spinach leaves
A block of Parmesan

The dressing
1 tsp honey
3 tbs olive oil
1 tbs balsamic vinegar
1 tsp finely chopped thyme
A couple of drops of truffle oil
Salt and ground black pepper
1/2 **tsp grain mustard**

Cut the aubergines in half lengthways, leaving the green stalk still on. Brush with olive oil, season with salt and pepper and bake in a hot oven (230°C/450°F/gas 8) for 10 to 15 minutes, and reserve.

Dry-fry the pine nuts for a couple of minutes over a moderate heat to brown them up, and reserve. Mix all the ingredients for the dressing together using a baby whisk or fork. Brush the Caesar's Mushrooms with a little olive oil, then season with salt, pepper and thyme. Grill for a minute or two on each side. Any excess juices can be poured into the dressing.

While the mushrooms are grilling, toss the leaves in most of the dressing and arrange on plates. Place the aubergine and mushrooms on the leaves, sprinkle with pine nuts, then shave Parmesan on top. Dribble the rest of the dressing on top, and serve immediately.

chapter 2

family life

'The humblest fungus betrays a life akin to our own. It is a successful poem in its kind.' (*Henry Thoreau, 1858*)

The answer to the riddle which so confounded our forebears, that of where exactly mushrooms *do* come from, is in its own way every bit as extraordinary as anything the ancients managed to dream up. Yet at one level the answer is very simple: mushrooms are the fruits by means of which a certain class of organism reproduces itself – just as apples and oranges are fruits of the class of organism we know as a tree.

It is when we look into the nature of the 'tree' of the piece that things start to get a little strange. For the creature in question, ubiquitous but usually invisible, is quite unlike anything we usually call by that name. Which is perhaps why it has no more folksy name in English than mycelium.

To understand the bizarre, almost sci-fi world of the mycelium, it makes sense to start with the smallest unit of fungal life: the spore. These tiny 'seeds' are produced in prodigious numbers (into the trillions in the case of the Giant Puffball), and are generally dispersed by one of two methods. In the majority of species, spores are borne on the wind, or even, because of their size, on the gentlest of air currents. A few significant others, including the Truffles and the Morels, take a more active approach, forcibly ejecting their spores when they sense that conditions are right. It is this distinction between 'spore droppers' and 'spore shooters' that forms the basis of the most important classificatory divide in the world of the large fungi.

If a spore is lucky enough to land in a suitable environment, it will germinate, producing a threadlike hypha, or filament. As the hypha grows through breaking down and absorbing nutrients from its surroundings, it tends to branch out, forming a mycelium. The individual mycelial networks of some species can grow to enormous dimensions. Single specimens of the *Armillaria* (Honey Fungus) family have been found which extend over hundreds of acres, making them by far the largest biological entities on the planet.

As the mycelia of some species of fungi grow, they exhaust the available nutrients in the soil and are forced to move ever outwards. It is this phenomenon that is responsible for the fairy rings which so spooked our ancestors. Some mighty tasty mushrooms grow in this manner, such as the eponymous Fairy Ring Champignon (*Marasmino oreades*), and the St George's Mushroom (*Calocybe gambosa*) which features in the recipes that follow. One ring of St George's known to us on Salisbury Plain is over a thousand years old and more than two miles (3km) in diameter.

Sorrel Soufflé with St George's Mushrooms

Serves 4

Sorrel and St George's Mushrooms tend to start protruding at the same time and often grow side by side in fields and pathways. This is therefore a marriage of convenience, as well as one of love. Don't be put off by the myth that soufflés are difficult – they are a piece of cake provided you don't blow it by prematurely opening the oven.

About 5 handfuls of fresh sorrel
75g/3oz butter
8 medium St George's Mushrooms
2 tbs plain flour
150ml/1/4 pint milk
100g/4oz mature Cheddar cheese, grated
4 eggs, whites separated from yolks
A pinch of ground nutmeg
Salt
Freshly ground black pepper

Reserve a few sorrel leaves for garnish, then blanch the rest in boiling water for 30 seconds, squeeze out the water, and finely chop. Reserve.

Preheat the oven to 200°C/400°F/gas 6 and grease a 1 litre/2 pint soufflé dish with a little of the butter.

Reserve the caps of 4 of the mushrooms and finely chop the rest. Heat another 40g/1 1/2oz of the butter in a thick-bottomed saucepan until it starts to bubble. Add the finely chopped St George's Mushrooms and fry for 5 minutes or so, until cooked.

Add the flour and beat into a paste with the butter and mushrooms. Cook this paste until it browns slightly. Whisk in the milk, adding it slowly to prevent lumps forming. Cook on gently for a couple of minutes, until the sauce thickens.

Remove from the heat and add the chopped sorrel, the cheese, egg yolks, nutmeg and salt and pepper to taste.

Place the egg whites in a large, grease-free mixing bowl (any grease will ruin the beating process that follows). Add a pinch of salt to stabilize the egg whites, and beat them with a hand whisk or electric mixer until they become stiff, with soft peaks which don't collapse under their own gravity.

Add a couple of spoonfuls of beaten egg white to the sauce and mix in with a rubber spatula. Fold the rest of the egg whites in gently, so as to keep as much air in the dish as possible.

Place this mixture into the soufflé dish and cook in the oven for half an hour. If you are not sure whether it is cooked, test it with a skewer. If it comes out clean, the soufflé is cooked. While the soufflé is in the oven, fry up the 4 remaining mushroom caps in the rest of the butter and season with a little salt and pepper.

Serve immediately in thick slices, with a St George's Mushroom cap and a garnish of sorrel leaves on each portion.

Pea Pancakes with Asparagus and St George's Mushrooms

Serves 4

A highly patriotic dish, although the mushrooms are more likely than domestic peas and asparagus to have appeared by St George's Day (23 April). But since they often fruit well into May and even later, your peas may well have time to catch up.

The pancakes
225g/8oz fresh or frozen peas
100g/4oz unsalted butter
2 eggs
100ml/4fl oz double cream
75g/3oz plain flour
Salt and ground black pepper

The Hollandaise
100g/4oz butter
3 egg yolks
A pinch or two of salt
1 tsp English mustard
Juice of half a lemon
1 tsp chopped dill

The rest
8 medium St George's Mushrooms, stalks trimmed
Salt, pepper, paprika
50g/2oz butter
Around 20 asparagus spears, trimmed
A sprig or two of flat-leaf parsley

To make the pancakes, first cook the peas. If they are frozen, just blanch for a minute. If they are fresh, cook for a little longer.

Melt just over half of the butter, pour into a food processor, add the peas and the rest of the ingredients for the pancakes and process until very smooth.

Reserve the mixture until needed, then fry 4 thickish discs in a little of the butter, turning each one when slightly browned, after a minute or so.

To make the Hollandaise, melt the butter in a pan, taking care not to burn it.

Place the egg yolks, salt, mustard and lemon juice in a blender, turn it on and gradually feed in the melted butter. This process thickens the sauce. After it is blended check the seasoning. Add the dill and stir in by hand.

Next, season the St George's Mushrooms with a little salt, pepper and paprika and fry gently in the butter for a few minutes on each side. Slice up 4 of them.

While you are doing this, you can be boiling the asparagus. Around 5 minutes should do. All the ingredients are now ready.

Make sure the plates are warm and cover each one in Hollandaise sauce, then line up 4–5 asparagus spears, facing the same direction. Place a pea pancake on top, then a St George's Mushroom cap, and finally garnish with the mushroom slices and parsley.

Returning to our newly formed hypha, we will find that its habit of branching out dramatically increases its chances of running into another hypha of the same species. When this happens, provided the hyphae are of different sexes (and the concept rather breaks down in the world of mushrooms, where some varieties have as many as 28,000 distinguishable 'sexes'), they may fuse to produce cells containing one nucleus from each 'parent'.

These two-nucleus or dikaryon cells will carry on behaving like normal hyphae until a particular series of environmental buttons are pushed telling the mycelium it is time to fruit. At this point, groups of filaments will start to clump together forming a dense, apparently chaotic tangle, which then gradually organizes itself into a tiny but discernible mushroom shape. This proto-mushroom then waits like a deflated balloon for its cue to expand. In many species, the trigger will be a shower of rain. In others it may be a shock to the system, like the sudden exhaustion of a food source. Either way, the expansion may be extraordinarily rapid. No wonder our ancestors were alarmed. They weren't to know that the mycelium had lined everything up behind the scenes.

The experienced mushroom hunter, of course, gets to know under what climatic conditions different species of mushroom are likely to fruit. But equally importantly, he or she needs to know where to look. This is where a second fundamental distinction between mushroom types comes into play, according to the kind of material, or substrate, on which they grow.

Of the three major groups into which mushrooms are classified according to how they grow, perhaps the least important from the point of view of edibility are the parasites, which obtain their nourishment from a living host, usually (but not always) a plant. One notable exception to the edibility rule is the Honey Mushroom (*Armillaria mellea*), although this is about the only good thing to be said for it. Gardeners dread the Honey Mushroom like nothing else. Able to spread over huge distances by means of sinister subterranean cords or rhizomorphs, it invariably destroys the trees on which it feeds and is almost impossible to kill. The only possible consolations for those in its grip are delicious dishes like the stir-fry that follows, although for the bereaved gardener they will probably turn to ashes in the mouth.

Pork and Honey Mushroom Stir-Fry

Serves 4

For a parasitic fungus, the Honey Mushroom sure can taste good. Be warned though: certain people – particularly gardeners – will not even set foot in a house where there is known to be Honey Fungus. Johnny's in-laws, for starters.

400g/14oz pork, cut into chunks
1 tsp ground cumin
1 tsp finely chopped garlic
1 tsp finely chopped thyme
1 tbs Worcestershire sauce
1 tsp tabasco sauce
1 tbs chopped oregano
1 tbs balsamic vinegar
1 tsp ground allspice
1 chipotle chilli (a Mexican smoked chilli), de-seeded and soaked in hot water for 20 min, then finely chopped
300g/10oz Honey Fungus caps
4 tomatoes
1 red pepper, roughly diced
1 yellow pepper, roughly diced
2 tbs olive oil
1 avocado, peeled and sliced and dribbled with the juice of a lemon

Marinate the pork for a couple of hours in a mixture of all the ingredients down to the chipotle chilli.

Blanch the Honey Mushrooms for 5 minutes in simmering water. Drain, discarding the water, and reserve.

Skin the tomatoes by scoring them with a sharp knife, then dunking in boiling water for 10 seconds. Plunge immediately into freezing cold water and the skin should fall off. Cut the tomatoes through the middle and squeeze out the seeds. Roughly chop the flesh.

Fry the pork and the peppers in the olive oil for 5 minutes in a large pan over a moderate heat, stirring continually. Make sure all the marinade juices go in as well.

Add the tomatoes and Honey Fungus and continue to cook for a further 10 minutes. Lastly add the chopped avocado and serve immediately on a bed of rice.

A much more benign set of mushrooms from the human perspective are the saprophytes, which live on dead organic matter, such as tree-stumps and compost. These fungi not only play a vital role in the natural recycling of nutrients, they are also particularly amenable to artificial cultivation. Shiitake, White Button Mushrooms and Oyster Mushrooms (respectively *Lentinula*, *Agaricus* and *Pleurotus* species) are all saprophytic fungi.

The real culinary prizes, however, tend to go to those fungi that live in symbiosis (mutual interdependence) with the roots of trees or bushes. These are the mycorrhizal fungi, which include such gems as Ceps, Truffles and Chanterelles. In all these species, hyphae actually penetrate the roots of the partner plants. The fungi are able to extract vital nutrients from their allies, while the plants find their surface area for taking up nutrients multiplied many times over, and everyone is happy.

It has been estimated that 90 per cent of all plants live in a mycorrhizal relationship with one or more species of fungus. It is precisely because these species are so inseparably bound up with their plant partners that they are so resistant to cultivation. And it is this essential wildness which gives them their romance and their glamour. It also does no harm at all to their prices.

We now know what our predecessors never could: that bizarre as it may seem, mushrooms are brought about by that unprepossessing white fluff that is occasionally revealed where the ground is disturbed, or when a piece of bark is stripped from a tree. The truth is that just beneath the surface, much of the Earth is permeated by fungal mycelia in almost incredible profusion. Studies have revealed that one cubic inch of soil can contain more than a million individual fungi. Their constituent hyphae may have a combined length of over a mile.

Taken together, the world's mycelia form a kind of enormous and intelligent natural internet, a network which connects almost all living things. In this sense, our ancestors were right all along. Mushrooms really are embodiments of the spirits of the woods and fields in which they grow.

The next question to consider is why mushrooms come in such a variety of shapes and sizes. The short answer is that all the diverse forms of fungi represent different strategies for disseminating spores as widely and successfully as possible.

The most familiar form of fruiting body is, of course, the capped mushroom, in which spores develop under the protection of a dome or canopy. This protective function is elegantly demonstrated by some species of the *Amanita* family, which will right their caps to the horizontal if disturbed, even after they are cut. Most but not all fungal caps are broadly umbrella-shaped, although in the case of the Chanterelle family the umbrellas would have to have been blown inside out.

Within this group, two major methods of spore production have evolved, both designed to maximize the surface area where this process can take place. The first, employed for example by the Agarics, including the common Field Mushroom (*Agaricus campestris*), takes place on the organs known to us as gills. The second, meanwhile, occurs within specialized pipes or 'tubules', which open out on the underside of the caps of the species in question as pores, giving them a characteristic spongy appearance. These are the Boletes, which include among their number the Cep (*Boletus edulis*), arguably the greatest mushroom of them all, and its close cousin the Bay Bolere (*B. badius*).

Whichever production technique the capped mushrooms use, they disperse their spores simply by dropping them when stimulated to do so by environmental conditions. But as we mentioned earlier, there is another group of fruiting bodies that takes an altogether less passive approach. These are the spore shooters, typified by the delectable Morel (*Morchella* species).

The strange, convoluted surface of the Morel starts to make sense when we consider what would happen to the spores of gilled and tubule mushrooms if they were ejected rather than dropped. The answer is that they would slam into the gill or tubule wall opposite, and not get very far at all. The different angles provided by the Morel's pits and ridges ensure that this does not happen, and that some of the spores shoot free.

When the appropriate conditions are met (perhaps after a shower of rain), a ripe Morel will fire out millions of spores simultaneously, utilizing a latent hydraulic pressure within its cells. If the light is right, this discharge can be seen as a puff of brownish smoke. Still more remarkably, it can be heard. The sound has been likened to the hiss of thousands of tiny aerosols all spraying at once.

Other ingenious spore dispersal techniques have been developed by edible fungi. Some, such as the appropriately named Hedgehog Mushroom (*Hydnum repandum*), use 'teeth' or spines in place of the gills and tubules employed by their relatives. The Puffball family (the gasteromycetes) develop leathery skins as they ripen, which eventually rupture to release volcanic clouds of powdery spores. Some even position themselves where the process is likely to be given a boost by their being trodden on by large hoofed mammals. It is with a recipe featuring one of these – the Giant Puffball (*Langermannia gigantea*), the greatest spore producer of them all – that we bring this chapter to a close.

'Fairy Ring Champignons can grow in large numbers, so bring a pair of scissors and a couple of bags if you are heading for a likely spot.'

Fairy Ring Champignons with Wild Asparagus

Serves 4

Fairy Ring Champignons start to appear during the late spring or early summer, typically growing on lawns or pastureland in fairy ring formation (as their name might lead you to expect). The grass on the inside of these pretty circles may be slightly faded, reflecting the fact that the growing mycelium will have exhausted the nutrients in the soil there before moving on outwards.

Delicate, fragile and tasty, Fairy Ring Champignons can grow in large numbers, so bring a pair of scissors and a couple of bags if you are heading for a likely spot. Fortunately, they dry rather well.

Wild asparagus has a mild grassy flavour, which isn't surprising, because a grass is exactly what it is. It shares the same season as the Champignon but can be quite hard to find, unless you know of a patch. If you can't get it, use regular asparagus or fine French beans instead.

A small bunch of basil, stalks removed and finely shredded
3 tbs good olive oil
A large bunch of wild asparagus (about 100 strands – they're very thin)
A medium-sized pile of Fairy Ring Champignons, checked for vermin and brushed free of dirt.
1 tbs butter
salt and ground black pepper
75ml/3fl oz rich chicken or vegetable stock.

Put a large pan of water on the stove to boil.

In a small bowl, add the shredded basil to two tablespoons of the olive oil and stir together. Mix in a little salt and pepper.

Fry the Champignons with the remaining olive oil and the butter over a moderate heat until they have released their juices and look glossy and cooked. Then add the stock, a little salt and pepper and reduce until most of the stock has evaporated. Reserve.

Simmer the asparagus in the water for about 8 minutes, then drain and season with a little salt and pepper.

Curl the wild asparagus round the mushrooms on a plate garnished with the olive oil and basil infusion.

Ceps and Pork Belly with Truffled Mash and Spinach

Serves 4

For this you need some pristine medium to large Cep caps. (We know where they grow, but we will never, never tell . . . never! HA HA HA!)

500g/1lb pork belly coated in 1 tsp paprika, 1 tsp soy sauce and 1/2 tsp salt
4 medium to large Cep caps, cut in half lengthways
Salt
50ml/2fl oz olive oil
800g/1 3/4lb large potatoes, cut into slices and boiled until soft
50g/2oz grated Gruyère cheese
100ml/4fl oz double cream
2 tsp white truffle oil
1 1/2 tbs chopped chives
50g/2oz butter
White pepper
400g/14oz baby leaf spinach

Bake the pork belly skin-side up in the oven at 220°C/425°F/gas 7 for about 40 minutes. Finish off under a high grill for a few minutes, or until the skin starts to bubble up. Cut into thick slices and keep warm.

Sprinkle the Ceps with salt and a little olive oil and bake in the oven for 10 minutes at the same temperature as before.

Mash the potato with the cheese, cream, truffle oil, 1 tablespoon of the chives, adding butter, pepper and salt to taste.

Bring a pot of water to the boil and briefly blanch the spinach. Squeeze the water out and toss in a little butter, salt and pepper.

Serve on hot plates, sprinkling the remaining chopped chives over the potato.

Bay Bolete with Red Wine and Venison

Serves 4

For some reason the Bay Bolete (*Boletus badius*) doesn't seem to be as sought after as its famous cousin, the Cep. Maybe this is because they tend to stain dark blue where handled and upon cutting, particularly as they get older. As soon as the Ceps have finished, this delicious mushroom comes into its own. It is resilient enough to survive the first couple of frosts of the year, so in Britain the season can run on well into November. The Bay Bolete is even better camouflaged than the Cep, with its dark brown cap merging in with soil and leaf litter.

In this recipe, the mushrooms take on the scented flavour of the marinade and soak up the meaty stock. The slow cooking won't affect their firm texture.

'As soon as the Ceps have finished, the delicious Bay Bolete comes into its own.'

4 x 150 to 200g/5 to 7oz slices of venison
2 tbs balsamic vinegar
200ml/7fl oz red wine
1 tsp orange zest
8 juniper berries, crushed in a pestle and mortar or blended in a coffee grinder
50g/2oz flour seasoned with a little salt and pepper
50ml/2fl oz olive oil
100g/4oz thinly sliced shallots
1 stick of celery, finely sliced
250g/9oz button Bay Boletes, trimmed and halved
200ml/7fl oz game or chicken stock
2 bay leaves
A sprig of thyme
4 cloves
2 tbs finely chopped parsley
Freshly ground black pepper and salt

Marinate the venison in the balsamic vinegar, red wine, orange zest and juniper for at least 4 hours, then pat dry and pull through the seasoned flour so that the steaks are evenly covered. Reserve the remaining marinade.

Heat up the oven to 170°C/325°F/gas 3.
In a large thick-bottomed ovenproof pan, heat up half the olive oil, then fry the venison steaks for a couple of minutes on each side over a moderate heat. Take out of the pan and reserve.

In the remaining olive oil fry the shallots, celery and Bay Boletes over quite a low heat, but with the lid on, stirring occasionally. After 15 minutes they should be quite soft.

Replace the venison, then add the rest of the ingredients apart from the parsley and the seasoning, and place in the oven for at least 2 hours, or until the meat is tender. The lid should be left on.

Before serving add the parsley, then season to taste. This dish is delicious with mashed potatoes and savoy cabbage.

Morels stuffed with Spinach, Ricotta, Parmesan and Pine Nuts

Serves 4

This mild, nutty dish of stuffed Morels on a pool of spinach sauce makes one of the most glamorous starters imaginable. Since the Morels are used whole, check them thoroughly for insects and dirt!

16 to 20 medium Morels
400g/14oz spinach blanched for 2 minutes in a large pan of boiling water, then cooled under the cold tap, squeezed out over a colander and finely chopped
50ml/2fl oz double cream
100ml/4fl oz chicken stock
Salt and ground black pepper
Nutmeg
100g/4oz leek julienne (cut into very thin matchsticks)
2 cloves garlic, finely chopped
50ml/2fl oz olive oil
250g/10oz Ricotta
40g/1½oz grated Parmesan
2 heaped tbs pine nuts, toasted in a pan for a few minutes under moderate heat until slightly browned
1 tbs chopped parsley

Trim off the Morel stalks flush with the head, leaving an opening into the cavity.

For the sauce, take a handful of the chopped spinach and place in the blender with the cream and chicken stock. (All the ingredients should be cold.) Add a little salt and pepper, grate in a little nutmeg and blend until smooth. Reserve in a small saucepan.

Fry the leeks and garlic in half the olive oil over quite a low heat for 5 to 10 minutes, until soft. Place in a medium-sized bowl and let cool. Combine the leeks with the Ricotta, Parmesan, most of the pine nuts, salt, pepper, the remainder of the spinach and the parsley.

Carefully spoon or pipe the mixture into the Morels, then place them on a baking tray, paint with the remainder of the olive oil, season and cook in the oven at 220°C/425°F/gas 7 for 15 to 20 minutes. Heat up the spinach sauce and spread on warmed plates. Place 4 to 5 Morels on top and garnish with the remainder of the pine nuts.

Fried Puffballs on Sourdough Bread

Serves 4

A field sprouting Giant Puffballs is quite a sight, as is a plate of glistening, milky-white Puffball slices on well-buttered sourdough bread. A few rashers of streaky bacon and some nice ripe tomatoes will finish this off nicely.

1 small firm Puffball, with no trace of discolouration
1 medium crusty loaf of sourdough bread
12 rashers of streaky bacon
2 beefsteak tomatoes, or 8 medium tomatoes, cut in half
A full butter dish
Salt and black pepper

Slice the Puffball into slices about 1cm/½in thick. They tend to squeak as you cut into them.

Slice the loaf up ready for the toaster. Put the grill on medium and gently cook the bacon until it is slightly crispy.

Season the tomatoes and fry them up in a little butter for a few minutes.

Spread soft butter on each side of the Puffball slices, season with a little salt and pepper, and fry them individually in a large pan until browned.

Toast the sourdough, butter each slice, then serve the Puffball slices on the toast with the bacon and tomato on top.

'A field sprouting Giant Puffballs is quite a sight. . .'

chapter 3

the wild heavyweights

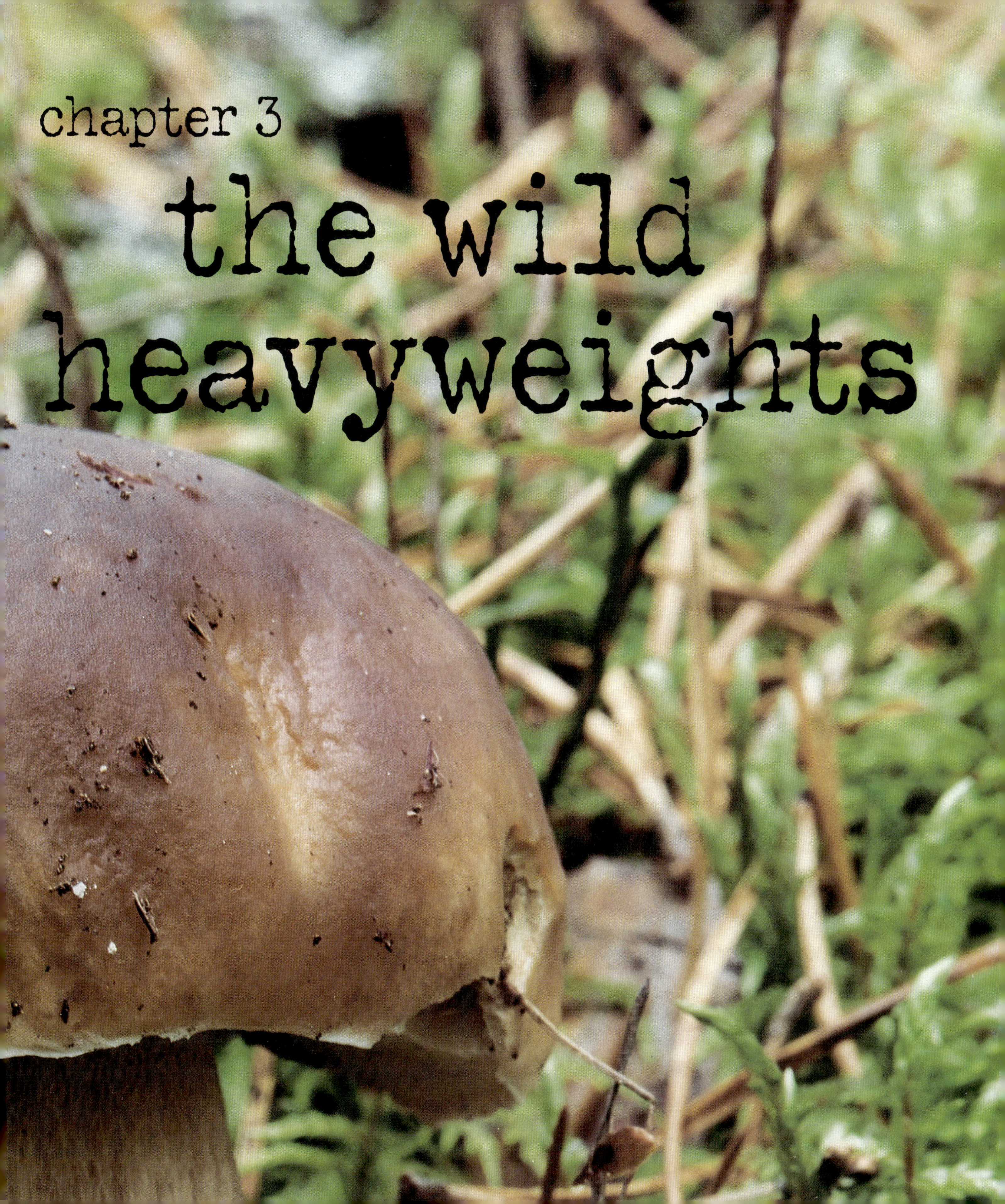

Sunday, very early. Weather humid and close. The sounds of the city gradually receded. We drove along a straight road for a long time. My nose told me we passed pig farms, pine trees and spilled diesel.

Eventually the car slowed down. Then potholes and the sound of a gate. The air was filled with heady aromas. The driver spoke quietly into his phone. Five minutes later, footsteps approached. A conversation ensued. A man with a strange accent led me into the woods. I felt hot, disorientated and thirsty.

'Take the blindfold off.'

I did as I was told, and was immediately blinded by the light. I took my jersey off. My guide shook my hand.

'Jersey,' he said.

'I know,' I said, slightly irritated. 'I put it on this morning.'

'No, it's my name, Jerzy. I am Polish,' he explained. 'Come now, the time has arrived. Please, feel the moss.'

I bent down and felt. It was damp and springy. Underneath were hidden lumps, like soft eggs. The sensation was thrilling.

Jerzy squatted down, and pushed his fingers into the moss. There was a sudden snap, and he emerged clutching a perfect baby mushroom: well rounded, the cap hardly distinguishable from the stalk. It was a pristine young Cep. Jerzy grinned a toothless grin. 'Borowik!' he beamed.

People have been collecting mushrooms for millennia. In 1991 hikers in the Italian Alps stumbled across the remains of a man clutching a primitive knapsack. Its contents included, in addition to one unidentified mushroom, a string of dried Birch Polypores. This multipurpose fungus can be used, among other things, as tinder for making fires, for the treatment of wounds or as the basis of an immune-boosting tea. Exactly what this individual had planned to do with it will never be known. It turned out that he'd died in about 3300 BC.

Although concrete evidence for the human use of mushrooms doesn't stretch back much further than our Italian friend, and although it may have been their hallucinogenic and aphrodisiac qualities that first moved people actually to write about them, it is a fair bet that our ancestors had been using them for one thing since the year dot: food.

One piece of evidence that points in this direction is the fact that some of our closest relatives relish wild mushrooms. Dian Fossey has established that mountain gorillas are particularly fond of the bracket fungus *Ganoderma applanatum*, while Nick is convinced he has discovered an appetite for Morels among Indian Langur monkeys (see the recipe for Morel Soup on page 48). Meanwhile, there is plenty of evidence of a love of mushrooms among other mammals. Truffles, for instance, are dependent on it: unless they are unearthed by a hungry (or randy) animal, they are utterly unable to reproduce.

Above all, our distant forebears are likely to have found mushrooms simply too good a source of free protein to turn down. This would have been particularly true in areas with severe winters, where the inhabitants would have had a powerful incentive to gather in everything edible before the big freeze (in Russia, for instance). It would also apply where the scarcity of useable agricultural land meant people had to be resourceful about what they ate (in Japan, for example). Maybe the traditional wariness some nations have of wild mushrooms is actually a kind of misguided luxury . . .

The reality is that of the few thousand identified species of large fungi, about thirty are very good to eat, and a similar number are downright poisonous. The remainder lie somewhere on a spectrum of scarcely worth bothering with, via indigestible, to utterly disgusting, but are unlikely to do you much serious harm. Among them are the armies of what mycophiles affectionately dismiss as LBMs ('Little

Brown Mushrooms'). Fortunately, the good mushrooms, though small in number, are all reasonably common. This chapter focuses on a fairly arbitrary Big Nine. It would have been ten, but on balance we decided to put the Field Mushroom (*Agaricus campestris*) in our next chapter, 'The Taming of the Shroom'.

Lest urban readers start to feel left out, we should point out that we have found examples of all the species featured in this chapter growing wild in central London. We were going to add 'with the exception of the Morel', but then miraculously, on the very day we handed the manuscript over to the publishers, we found them growing in profusion on a building site near Heathrow airport. The happy truth is that, though they may be scarce, there are delicious, edible wild mushrooms to be found pretty much everywhere, provided you know where and when to look. From Truffles in the deserts to Ceps on the margins of the Greenland ice-cap. And perhaps even under our very noses.

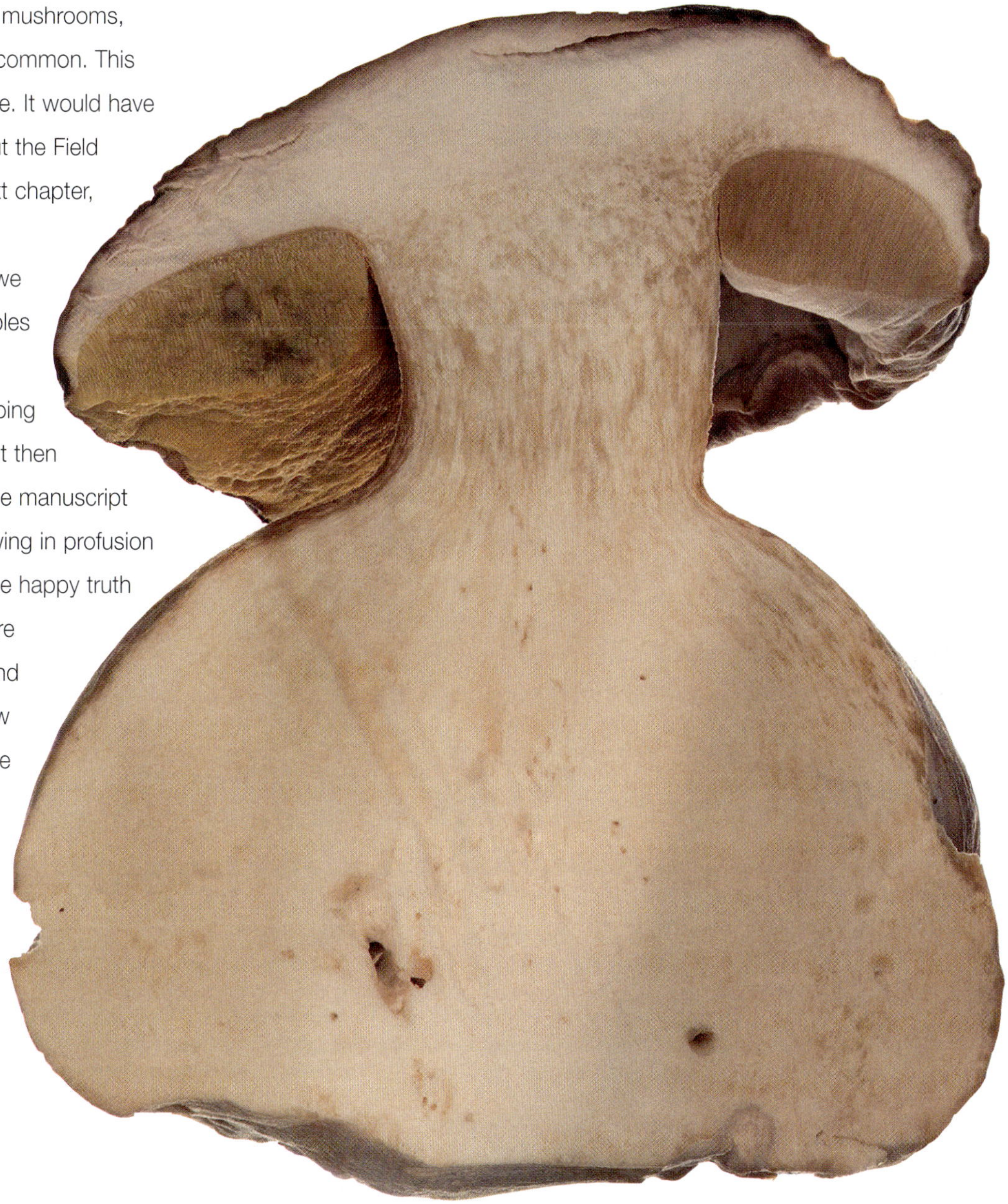

'... edible wild mushrooms are to be found pretty much everywhere, provided you know where and when to look.'

'. . . there might be girolles around here.'

Mushroom hunting awakens instincts most of us had forgotten we possessed. Our vision and sense of smell sharpen and the adrenaline starts to flow. We start to find ourselves mysteriously drawn to look in one place rather than another by a force we cannot begin to understand, but whose tug is irresistible. More often than not, it is right. In foraging for mushrooms, we plug straight into our Inner Hunter-gatherer.

Although some cultures are undoubtedly more attuned to searching for fungi than others, we reckon the mushrooming instinct is innate. All that is needed is to find a way to access it. There is a lot to be said for strategic hunting – that is to say, going by the book in terms of where you look for what (particularly with mushrooms that are linked with particular tree species), and when. But the best method is simply to get into the habit of walking around with this one thought firmly in mind: 'Now, where are the mushrooms?' Because you can bet your bottom dollar they're out there somewhere.

As time goes by, you will start to get a feel of what species tend to grow where, and in what conditions. The accent should be on the 'tend', as mushrooms are far too free-spirited always to obey the rules. But you will gradually start to catch yourself thinking things like 'mmm, this could be Morel territory', or 'there might be girolles around here.' This means you are starting to develop the all-important sixth sense.

Given that almost all cultures have a wealth of fungal goodies at their fingertips, it seems all the more peculiar that some exploit it gleefully while others run screaming from anything that isn't your basic Supermarket White. Particularly as there are just as many poisonous varieties in Warsaw as in Walsall. Maybe it is all a matter of national temperament. Whatever the reason, the upshot is that many people – particularly in Britain and Ireland – don't know what they are missing.

Not that there is nothing to be said for a bit of Anglo-Saxon caution. Poisonings really do happen: by the end of July 2000, they had already accounted for 111 Ukrainians and 73 Russians. But, fortunately, many of the best edible species are relatively easy to identify, provided you are scrupulous about following the rules (see the 'Essentials' section at the end of the book). That said, there is never room for complacency where fungi are concerned. John Gerard, over four centuries ago, concluded his musings on mushrooms with a sobering warning which still applies today:

'I give my advice unto those who love such strange and new-fangled meates, to beware of licking honey among the thorns, lest the sweetness of the one do not countervaile the sharpness and pricking of the other.'
(Herbal, 1597)

ceps

(*Boletus edulis*)

Rainy weather would bring out these beautiful plants in profusion under the firs, birches and aspens in our park, especially in its older part, east of the carriage road that divided the park in two. Its shady recesses would then harbour that special boletic reek which makes a Russian's nostrils dilate – a dark, dank, satisfying blend of damp moss, rich earth, rotting leaves. But one had to poke and peer for a goodish while among the wet underwood before something really nice, such as a family of bonneted baby edulis or the marbled variety of scaber, could be discovered and carefully teased out of the soil.
(Vladimir Nabokov)

One of the greatest pleasures for a mushroom hunter is coming across a pristine group of Ceps, nestling in a patch of woodland grass like a batch of freshly baked loaves. Their appearance says only one thing: 'Eat me!'

Boletus edulis is quite simply one of the best mushrooms in the world. Throughout Europe, from late summer onwards, they are gathered in their millions. Those that are not eaten immediately are lovingly dried or pickled or frozen for posterity. Their popularity is such that in some countries legislation has had to be passed to prevent over-harvesting. In areas of Italy, for example, you need a permit to pick.

The popular names of the Cep around the world reflect the esteem in which it is held. The traditional British name is 'Penny Bun', while to Americans it is just 'King Bolete'. The Romans called Ceps *suilli*, the Latin for swine, and their Italian descendants know them as 'Porcini', or 'little pigs'. No one knows the reasoning behind the pig motif – perhaps it has something to do with the Cep's dense, almost meaty flesh. What it does do is tell us a great deal about the affection felt over the millennia towards this scrumptious mushroom.

The ideal preconditions for a bumper crop of Ceps are a hot, dry summer followed by a warm, wet, early autumn. We know of a Scottish enthusiast who during one such glut managed to pick over a ton in one twenty-four-hour period (and that was just the button ones), earning £8,000 for his trouble. Despite their substantial nature – a big one can weigh more than a kilo – the entire growing cycle is only about two weeks. Should you be lucky enough to find any, our advice is simple: get picking, and keep *stumm*. It also pays to remember where you found them. Ceps often fruit in the same spot year after year.

Various clues alert Cep enthusiasts to the possibility that their quarry may be nearby. Beech trees are definitely a promising sign. Fly Agarics can be another. Some believe that The Miller (*Clitopilus prunulus*), a small white mushroom with a strong smell of flour, is an almost infallible guide to the proximity of *Boleti*. But however you come to find them, just pray that the maggots haven't got there first. They love them almost as much as the Italians.

One of the great things about the Cep is how well it lends itself to being dried or frozen. Dried Ceps have a concentrated, musky flavour which is far richer than that of their natural state. Much of this flavour leaches into the water in which they are reconstituted. The result is an intense mushroomy stock which is a superb ingredient in its own right.

Roasted Vegetables with Ceps

Serves 4

This vegetarian dish is easy to make and combines many textures and tastes while preserving the integrity of each ingredient. The roasted pepper purée adds a welcome flash of sweetness. The vegetables used are some of our personal favourites, but feel free to experiment with your own. Serve with basmati rice.

Some vegetable oil for frying
1 medium leek, cut lengthways into very thin strips
A little plain flour
1 small butternut squash, cut in chunks
2 large red peppers, cut lengthways and de-cored
8 long (banana) shallots, or normal ones, peeled
2 small aubergines, cut lengthways
8 smallish Ceps, cleaned
8 baby carrots
4 large, fleshy tomatoes, cut in halves or quarters
Around 100ml/4fl oz olive oil
Coarse, flaky salt
Ground black pepper
A few sprigs of thyme (lemon thyme is very nice)
A squirt or two of balsamic vinegar

Heat the vegetable oil in a frying pan until shimmering-hot. Dust the leeks with flour and fry for 30 seconds or so, until crispy. Remove and drain on kitchen paper.

Place the squash, peppers, shallots, aubergines, Ceps, carrots and tomatoes in an oven dish and season with half the olive oil, salt, pepper, thyme and a little balsamic vinegar. Put in the oven preheated to 220°C/425°F/gas 7 and cook for 30 minutes or so. (Note that some of the vegetables may be ready earlier than others. If they look ready, take them out and store on top of the stove.)

Place the cooked red peppers in a food processor with 50ml/2fl oz olive oil, a squirt of balsamic vinegar and salt and pepper. Add the juices from the roasted vegetables and blend.

Make sure the plates are warm. Cover with roast pepper purée, arrange the vegetables on top and sprinkle with the fried leek.

Ceps with Pasta

Serves 4

If you are lucky enough to find fresh, maggot-free Ceps and want a fine way of enjoying them, try them with pasta and a little cream. The textures are wonderfully complementary. For this recipe, only firm young specimens will do.

A small, whole head of garlic
50ml/2fl oz olive oil
A handful of pine nuts
1 shallot, finely chopped
4–6 small firm Ceps, sliced
100ml/4fl oz double cream
Salt and black pepper
A small bunch of flat-leaf parsley, chopped
800g/1½lb fresh pasta

Slice the top off the head of garlic, drizzle with olive oil and wrap in silver foil. Bake in the oven at 220°C/425°F/gas 7 for half an hour. Leave to cool, then squeeze out the flesh and discard the skin. Reserve.

Dry-fry the pine nuts for a minute or two, until slightly browned. Reserve these also.

Heat up the remaining olive oil in a pan and fry off the shallot over a medium heat for a couple of minutes.

Add the Ceps and fry on for 5 minutes, until they have released a little moisture and it has evaporated off.

Add the pine nuts, garlic, double cream, salt and pepper and the parsley. Bring to the boil and reserve.

Meanwhile, cook the pasta al dente, add the sauce and serve.

Beef Wellington with Ceps

Serves 6

This dish apparently derives its name from the Duke of Wellington's legendarily shiny boots rather than because he had any particular love for it. In fact, he was a vegetarian. Traditionally, Button Mushrooms are used in the filling, but fresh Ceps make for a luxurious variant. You could use dried ones at a pinch, but if you do, make sure they are of a high quality and that you soak them well in warm water before use.

600g/1¼lb puff pastry
3 shallots, finely diced
2 cloves garlic, finely chopped
50g/2oz butter
250g/9oz fresh Ceps, chopped into duxelles
A sprig or two of thyme, leaves stripped off stalks
1 tbs chopped parsley
Salt and ground black pepper
6 x 150g/6oz fillet steaks
1 tbs olive oil
180g/7oz foie gras pâté
1 egg, beaten, with a dash of water added

Roll the puff pastry into 6 rectangles, each large enough to wrap around one steak, and put aside. Preheat the oven to about 225°C/440°F/gas 7–8.

Fry the shallots and garlic in the butter over a moderate heat until soft, about 3–5 minutes. Add the Ceps, thyme and parsley and a little salt and pepper. Cook until the Ceps have released their juices and these have evaporated somewhat.

Season the steaks with a fair amount of pepper and a little salt. Fry them off in olive oil in a thick-bottomed pan over quite a fierce heat for a minute or two on each side to seal the juices in. (Fry them for a little longer if you like your steak well done.)

Lay out the pastry sheets on a floured surface and place a little mushroom mix on each one. Top with a steak and spread with one-sixth of the foie gras pâté.

Paint the edges of each sheet of pastry with a thin layer of egg wash, then bring up the sides and neatly press together. Glaze each Wellington with egg wash. Add some shaped pastry trimmings for decoration and paint them on with the egg if you feel creative.
Cook in the oven for around 20 minutes, until nicely golden, and serve.

Borscht with Dried Ceps

Serves 4

Every now and again, a year comes around where conditions are perfect for Ceps. They may then spring up in enormous numbers – particularly, for some reason, in conifer plantations. We know of one spruce plantation where, in an exceptional year, a friend put his team of pickers to work with spotlights on their heads, so they could harvest right around the clock. They still had to leave most of the mushrooms there, because there were just too many of them. Fortunately, as the Slavs and the Italians have long known, Ceps dry extremely well in slices. Nick dries them on a rack over his Aga, but you could use a radiator or a slightly warmed oven. As they dry, the Ceps will fill the house with the most entrancing aroma. Once they are done, they will last for years, provided they are properly stored in dry, airtight containers.

1.5 litres/3 pints beef stock
A handful of dried Ceps
50g/2oz butter
2 cloves garlic, chopped
100g/4oz sliced carrot
125g/5oz sliced onion
500g/1lb beetroot, peeled and sliced
300g/10oz potatoes, peeled and sliced
100ml/4fl oz tomato passata (sieved tomatoes)
3 tbs balsamic vinegar
A pinch of brown sugar
200ml/7fl oz sour cream
Salt and freshly ground black pepper
2 tbs chopped parsley

Heat up the beef stock to boiling point, take off the heat, and throw in the dried Ceps. Soak for 20 minutes.

Melt the butter in a thick-bottomed pan, add the garlic, carrot and onion, then remove the Ceps from the stock with a slotted spoon and add them to the frying vegetables. Continue to fry for 15 minutes, until soft. Pour in the beef stock slowly (taking care not include the dregs, which may contain grit from the mushrooms).

Add the sliced beetroot and potato, bring to the boil and simmer for 30 to 40 minutes, or until well cooked.

Add the passata, balsamic vinegar, sugar and sour cream, then blend using a stick blender. Add salt and pepper and the chopped parsley. This soup is delicious with thin slivers of deep-fried beetroot.

Warm Cep Salad with Artichoke and Wild Rocket

Serves 4

For this salad, we use wild rocket, which has a more pronounced taste than farmed varieties. As for the Ceps, it is particularly important to ensure they are free of insects, as in this recipe they are scarcely cooked.

Serve this dish in separate portions to bring out its considerable beauty.

4 globe artichokes
Juice of a lemon
10 walnut kernels
Vegetable or olive pommace oil for frying
A large bunch of fresh sage, leaves separated from stalks
Good-quality salt, such as fleur de sel de Carmargue
4 medium Ceps, sliced thinly
4 tbs olive oil
Freshly ground black pepper
4 handfuls of rocket
1 tbs balsamic vinegar
Generous shavings of Parmesan

Boil the artichokes in a large pan of slightly salted water for 45 minutes, then plunge them in cold water. Peel the leaves from the artichokes, then pull out the hairy choke, leaving the heart intact. Dribble a dash of lemon juice on top to prevent them from going black.

Bake the walnuts at 200°C/400°F/gas 6 for 5 minutes, or until slightly browned. Crumble.

Heat up the vegetable/pommace oil in a pan until a test sage leaf thrown in fizzes up. Then fry the sage leaves off in 2 or 3 batches. They are ready when they stop sizzling. Drain on kitchen paper and sprinkle with a few grains of salt.

Fry the Ceps in 2 tablespoons of olive oil for a couple of minutes on a high flame. Sprinkle with a little salt and black pepper. Keep warm. Toss the rocket with the rest of the olive oil, the balsamic vinegar and the lemon juice.

Arrange the salad on 4 plates with an artichoke heart in the middle. Surround with the warm Cep slices, top with sage leaves, the crumbled walnuts and some Parmesan shavings, and serve.

morels

(*Morchella esculenta*)

The Morel arouses strange passions in those who fall under its spell. 'All's fair in love, war and Morel hunting,' as one devotee bluntly put it. More than one murder has been attributed to lust for this strange, brain-like fungus. Part, but only a part, of its attraction can be put down to the fact that, unlike most mushrooms, it appears in the spring.

Nowhere is Morel mania more highly developed than in the Midwest of the United States. The rumblings are felt as early as late February, when reports come in of the year's first finds down in the Mississippi delta. But it is in May that the whole thing really explodes, particularly (to continue with the alliteration) in Michigan. Whole communities are possessed by the Morel bug. There are mass forays, competitions and numerous festivals. There are even Morel bulletins on the radio.

One of the many things which makes the Morel such an object of fascination is its legendary elusiveness. Perhaps because it tastes so good, it has been forced to evolve a devilishly effective camouflage. Yet once penetrated, this cloak of invisibility can dissolve miraculously. Frequently, on a Morel hunt, nobody will find anything for hours. Then someone will spot one, and suddenly they'll start to appear everywhere. It is almost an act of faith. Experienced Morel hunters are rather like residents of the rainforest, who are able to pick out fish in what outsiders can only see as raging torrents of water.

Specifying where Morels are likely to be found is extremely difficult, because they are capricious even by mushroom standards about where they will and will not grow. However, there are certain discernible patterns. For one thing, Morels are known to be particularly fond of burned and recently disturbed ground. This was poignantly illustrated in the spring of 1919, when a bumper crop of Morels sprang up all along the former No Man's Land, the shell-blasted strip between the rival trenches of World War One. The same tendency has sometimes led overzealous individuals in Switzerland and Italy to deliberately set fires to promote the growth of their favourite mushroom. On a less dramatic level, we have a friend whose speciality is finding Morels in construction sites, particularly ones for out-of-town supermarkets.

What Morels really like are temperate forests, often at high altitudes in areas with continental climates. Rather surprisingly, India and Pakistan are each responsible for about a third of the world's commercial crop. In the latter at least, this may have something to do with the fact that the locals tend not to eat them, so the majority of the Morels make it through to export. This would never happen in the Midwest.

Other items of Morel trivia include the fact that Louis XIII of France apparently died while stringing a basket-full for drying purposes, and that in Nepal they dry them over heaps of animal dung. But what really matters with Morels is how they do as food. Nutty, earthy and perhaps a little smoky too, they also have a texture and visual appeal to take them right to the top of the culinary tree. Their pitted surface makes them perfect absorbers of creamy sauces, and their hollow stalks are an open invitation to stuff. Morels, if you didn't catch the message, are very, very good.

For centuries, the cultivation of Morels was beyond our grasp. Recently, however, there are signs that all that may be about to change. During the mid-1980s, a cultivation technique was successfully patented in the United States, and although it has yet to be refined to the point where large-scale production is really commercially viable, some artificially grown Morels have already come to market. Some reviewers have claimed, perhaps inevitably, that they lack the full flavour of their wild cousins.

Morels, like Ceps, dry exceptionally well. Once again, the water they soak in becomes at least as important a flavouring agent as the mushroom tissue itself.

Morel Soup

Serves 4

A couple of years ago, Nick was on a trekking holiday in the Parvatti Valley in India when he came across a wizened saddhu or wandering mystic wearing what was unmistakably a dried Morel around his neck. Ever the mycological opportunist, he began to follow the old man around. Eventually, he caught the saddhu's attention, and pointed imploringly at the shrivelled fungus. 'Ah . . . Goochi!' said the holy man. This turned out not to be an ironic remark about the lowly status of his fashion accessory, but instead the local name for the Morel. Within an hour, the saddhu had rustled up a small pile of Goochis, which we later learned play an important role in traditional Ayurvedic medicine.

With his appetite thus whetted, Nick couldn't restrain himself from setting off into the forests surrounding the hot spring of Kirganga, approximately 3,500 metres (11,500 feet) above sea level. After a few hours, he heard the screeching of Langur monkeys up in the trees. This troupe led him deeper and deeper into the pine forest, until he found himself in a clearing where the soil had recently been disturbed. As his eyes adjusted, he realized he was surrounded by Morels. Some of them were as big as hand-grenades.

That night, he and his companions feasted on Morel soup, mopped up with parathas, the tasty local flatbread.

400g/14oz small, fresh Morels
50g/2oz unsalted butter
2 finely chopped shallots
1 litre/1½ pints chicken stock
250ml/½ pint double cream
6 egg yolks
Salt and freshly ground black pepper

Fry 8 of the Morels in half of the butter for 4 minutes over a moderate heat. Watch out: they spit! Reserve, leaving any excess butter in the pan.

Finely chop the remaining Morels, and fry in the rest of the butter along with the shallots. Add the chicken stock and bring to the boil. Take off the heat, pour in the double cream and whisk in the egg yolks with a little salt and pepper. Place back on the heat and slowly bring back up to temperature, whisking continuously. The soup will thicken slightly like a thin custard. As soon as it does, serve with the Morel garnish on top.

Braised Lamb with Morels and Madeira

Serves 4–6

This is an easy and impressive dish, in which the lamb is slow-braised with heaps of Morels and root vegetables. All you need to do during the cooking process is occasionally turn the lamb. The ideal cooking receptacle would be a large two-handled cast-iron pan with a heavy lid, perfect for the stove top where the vegetables are fried, and then for the oven where the lamb is gently roasted.

A leg of lamb
Salt and freshly ground black pepper
Olive oil
150g/6oz cubed pancetta
1 onion, diced
2 carrots, diced
2 sticks celery, sliced
4 cloves garlic, peeled
1 turnip, diced
400g/14oz fresh Morels
A couple of sprigs of thyme
2 bay leaves
2 sprigs of rosemary
200ml/7fl oz Madeira

Season the lamb with salt and pepper, heat the pan up with a little olive oil at the bottom, then lightly brown the lamb all over. Take the pan off the heat and remove the lamb.

Place the pancetta, onion, carrot, celery, garlic and turnip in the pan in a little extra olive oil and gently fry for 10 to 15 minutes. Add the Morels whole, and continue to fry for a further 5 to 10 minutes.

Add the herbs and Madeira and a drop of water. Place the lamb on top of the vegetable base, and gently push it down so that a few of the vegetables and some of the liquor rise up the sides. Sprinkle with a little salt and pepper and cover with the lid. Place in the oven at 170°C/325°F/gas 3 for 3 hours, turning the lamb every 40 minutes or so.

When the dish is ready take the lamb out, skim off any oil from the surface of the sauce, check for seasoning, and then serve the lamb with the sauce and vegetables.

This dish is perfect with new potatoes.

Morels with White Asparagus and Garlic Leaves

Serves 6

As luck would have it, a friend of ours has a company specializing in the importation of mushrooms and other wild foods, and it lies slap on the way to Nick's place of work. While most of us are still in bed, the two of them sit guzzling coffee and chatting about the specimens that come in, where they originate and the diverse characters involved in this colourful business. Nick gets first pick of newly arrived products, or at least he likes to think so.

Recently, on the suggestion of the proprietor, Nick picked up half a kilo of Morels, a handful of wild garlic leaves and two bunches of white asparagus. This is what he did with them when he got home.

A whole head of garlic
Olive oil
500ml/1 pint rich chicken stock
400g/14oz fresh Morels
75g/3oz unsalted butter
1 red onion, thinly sliced
250 ml/½ pint crème fraîche
800g/1¾lb white asparagus
A large handful of wild garlic leaves, torn into strips (if you can't find them use chives or suchlike)
Salt
Ground black pepper

Cut off the shoot end of the head of garlic, and discard. Sprinkle the remainder with olive oil, wrap in foil and roast at 220°C/425°F/gas 7 for about 30 minutes, or until soft. After it has cooled, squeeze out the pulp and reserve.

Pour the chicken stock into a saucepan and reduce by about half by evaporating on a slow boil.

Brush or wash the Morels clean, then cut them in two lengthways, remembering to do an insect audit.

Melt the butter in a thick-bottomed pan and stew the Morels slowly for at least 20 minutes. Then add the red onion and half the garlic and continue to cook for another 10 minutes. Add the cream, chicken stock and a little salt and continue to simmer until needed.

Bring a large pot of water to the boil. Peel the asparagus spears from close to the tips right down to the bases (their outer layer can be quite stringy), then chop off the bottom 2cm/1in or so and discard. Boil the asparagus gently for about 15 minutes, until soft.

Stir the wild garlic leaves into the Morel mixture, then serve immediately topped with the asparagus and a twist of black pepper.

Raclette with Morels

Serves 4

Raclette is a simple Swiss dish which derives its name from the French word *racler*, meaning 'to scrape'. It is also the name of the particular kind of cheese which is used to make it. In the past, it was prepared by placing half such a cheese in front of an open fire until the face had softened. The shepherd (the quintessential raclette eater) would then scrape off a thick slice and mix it with new potatoes and pickled vegetables.

Nowadays, special raclette grills and ovens are available, where you can melt the cheese in a thick-bottomed tray and grill complementary vegetables and meats at the same time. Unless you have one, you will need one-portion heatproof dishes for this recipe, as you will be shoving them under a hot grill.

Johnny's Swiss friend Catherine swears that in her native country you are not allowed to give blood within three days of eating raclette or cheese fondue. This is presumably because of temporarily stratospheric cholesterol levels. Don't be put off. This is a delicious treat, with the nuttiness of the Morels perfectly complementing a similar quality in the cheese.

200g/7oz fresh Morels, or a large handful of dried ones plus some chicken stock
400g/14oz new potatoes, unpeeled
1 large red onion, sliced thinly
50g/2oz unsalted butter
400g/14oz Raclette cheese, cut into 4 slices
Salt and ground black pepper
Sprigs of flat-leaf parsley

If you are using fresh Morels, clean them and cut in half lengthways. If using dried ones, soak them in hot chicken stock for half an hour, then drain. Reserve.

Boil the new potatoes in their skins for 20–30 minutes, until soft.

Gently stew the onion and Morels in the butter for around 10 minutes, until soft.

On each individual dish place a thick slice of cheese, 3 or 4 potatoes and 1 tablespoon of Morel and red onion mixture. Sprinkle with a little salt and a generous pinch of ground black pepper.

Melt under a hot grill until the cheese is molten and slightly browned. Serve immediately, garnished with the parsley.

'... you are not allowed to give blood within three days of eating raclette or cheese fondue...'

chanterelles

(*Cantharellus cibarius*)

If you are ever out in the woods and come across what seem to be the scattered contents of a packet of apricot-flavoured crisps, rejoice. You have almost certainly chanced on a clump of Chanterelles (*Cantharellus cibarius*), one of the precious jewels of the fungal crown.

Most of the species featured in this book blend in with their environments, at least to some degree. Chanterelles are different. They positively scream out their presence. This characteristic has led to them being compared to flowers, and indeed there is something floral about the Chanterelle. The sensation of finding a seam of them in the woods, for instance, is enchanting in much the same way as coming across a patch of bluebells. Only bluebells seldom make one start to salivate.

In Northern Europe, Chanterelles begin to grow in May or June, and carry on until the first frosts of autumn. In the off-season, they are imported from suitable areas of the Southern Hemisphere, such as the Highlands of eastern Zimbabwe. The Chanterelle is a truly global mushroom, with a distribution stretching from the Equator to the Poles.

Girolles, as they are also called (not least by the French) can lie dormant in button form for weeks, waiting for a suitable moment to expand to their full dimensions. Most chefs prefer them small, but Nick is a fan of the big ones. Girolles can be recognized by their lobed, irregularly shaped caps, by their egg-yolk colour and by the fact that their 'gills' continue to run down a portion of their stalks.

Chanterelles have a relatively short shelf life, owing to a high moisture content, but they keep in the fridge for up to a month. They are often dried, although preserved ones lack the succulence of their fresh siblings, a succulence which led one author, in 1755, to claim that Chanterelles could bring the dead back to life if only they were placed in their mouths.

When the Chanterelles start to die out, roughly as the first Blewits begin to appear, addicts can console themselves with the Winter Chanterelle or Yellowlegs (*Cantharellus tubiformis*, aka *C. infundibuliformis*). This reliable little fungus carries right on through to Christmas, and is usually found growing among pine needles. Another comforting thought is that the Chanterelle may be right on the verge of successful cultivation.

Chanterelle and Roasted Garlic Quiche

Serves 4

This quiche could be made with almost any mushroom – or any mixture thereof. We memorably made it with Caesar's Mushrooms when in the South of France; but it can hold its head up high whether it contains plebeian fungi like oyster mushrooms or aristocratic candidates like Chanterelles…

Pie Shell

100g/4oz plain flour
1/2 tsp salt
1 tsp sugar
8 tbs unsalted butter
3–4 tbs cold water

Filling

6 eggs
100g/4oz Emmenthal
100ml/4fl oz cream
1/2 tsp salt
1 head of garlic, roasted as in the recipe for Ceps with Pasta (see page 43) and then squeezed out
200g/7oz Chanterelles, torn into strips
Paprika

To make the pastry shell, begin by combining the flour, salt, sugar and butter in a food processor for 10 seconds or so until the mix has a crumbly texture. Then rub in the water with a spoon. Flatten the mixture, wrap it in cling film and refrigerate for 1 hour.

Roll out the pastry, transfer to a 20–25cm/8–10in pie dish, and shape round the edges. Prick the bottom with a fork. Line the pastry with baking paper and weight down with dried pulses or rice. Bake for 15 minutes in an oven preheated to 180°C/350°F/gas 4. Then remove the paper and the lentils or rice, and bake for a further 15 minutes at 200°C/400°F/gas 6.

The crust is now ready for the filling. To make this, first mix all the ingredients (including the squeezed garlic) – except the mushrooms and paprika – while warming the crust in the oven for 10 minutes. Pour in the mix and place the mushrooms on top, gently mashing them in slightly, but leaving some still protruding. Sprinkle with paprika.

Return to the oven, and bake for 30–40 minutes at 180°C/350°F/gas 4 before serving.

Puy Lentil Stew with Chanterelles

Serves 4

Puy lentils originate from the Auvergne in France, which is one of the least populous areas of the country. Fewer people naturally means there are more mushrooms to go round, and the Ponots (the local inhabitants) take full advantage.

This is a slow-cooked, deeply flavoursome stew, nutty and fragrant. It will keep in the fridge for days.

250g/9oz puy lentils
Just over 1 litre/1 3/4 pints chicken or pork stock
150g/5oz fresh pork belly, cut into lardons
A little pork fat or olive oil
150g/5oz sliced shallots
2 cloves garlic, chopped
200g/7oz small Chanterelles, brushed clean
2 bay leaves
2 sprigs of thyme
Salt and black pepper
100ml/4fl oz crème fraîche

Place the lentils in a pan with the stock and bring to the boil. Turn the heat down and simmer.

Fry the lardons separately over moderate heat with the olive oil or pork fat for 5 minutes or so until they have gained a little colour. Then, with a slotted spoon, put them in with the lentils, leaving the fat in the pan.

Fry the shallots, garlic and Chanterelles in the fat for up to 15 minutes over medium heat, until the mushrooms have released their juices and sucked them back in again. This process will improve their flavour immensely.

Add this mixture to the lentils, then add the herbs and a little seasoning.

Simmer the stew for 1 1/2 hours at least, checking frequently that it doesn't run dry. If the stock has evaporated too much, top up with water or stock to just above the lentils.

Serve with a generous blob of crème fraîche on each portion.

Winter Chanterelles with Roast Beef and Red Wine Jus

Serves 4

This is admittedly not the simplest recipe in the book (it's as well to begin making at least the stock the day before you plan to eat, and you will also have some left over for use in another dish), but it is one of the most rewarding. It makes a cracking Sunday lunch when the wind howls outside.

The beef stock

1 kg/2¼lb beef (or preferably veal) bones, cut into small pieces
1 large carrot
1 stick celery
1 onion
500g/1lb cheap cut of beef or veal
Vegetable oil
3 litres/5½ pints water
1 bouquet garni

The jus

6 shallots, unpeeled and roughly chopped
2 cloves garlic, unpeeled and crushed
50ml/2fl oz olive oil
50ml/2fl oz balsamic vinegar
100ml/4fl oz port
500ml/1 pint red wine
1 litre/1¾ pints of the beef stock
25g/1oz tomato purée
A sprig of parsley
2 bay leaves

The main dish

700g/1½lb fillet steak
Cracked black pepper
Salt
1 tbs olive oil
400g/14oz Winter Chanterelles, cleaned
50g/2oz butter
A sprig of tarragon

To make the stock, preheat the oven to 220°C/425°F/gas 7. Throw the bones, vegetables and meat into a roasting pan, baste with around 3 tablespoons of oil, and roast for 45 minutes.

Place the above items in a deep pan, remembering to scrape out the bottom of the roasting dish for added flavour. Add the water and bouquet garni and simmer for 4 to 6 hours.

Pass the stock through a fine sieve, then skim off the fat from the top. Continue to boil the liquid until you have reduced it by half.

To make the jus, fry the shallots and garlic in the olive oil on a medium heat. If they start to brown too much, turn the heat down. When they have softened (after about 15 minutes), add the vinegar and port. Turn the heat up and reduce by at least half.

Add the red wine and the stock and reduce again by at least two-thirds.

Add the remaining ingredients and simmer for 20 minutes.

Pass through a fine sieve or muslin, and store until needed in the fridge or freezer.

For the 'main event', begin by sprinkling the fillet steak with cracked pepper and a little salt. Then brown it in a pan in the olive oil to seal in the juices and bang it in a hot oven, 230°C/450°F/gas 8 for a quarter of an hour (less if you like it rare, more for well-done).

Fry the Winter Chanterelles in the butter. They will give off a lot of juices. Cook until these have been soaked up again.

Add the Winter Chanterelles to 500ml/1 pint of the jus and season to taste with salt. Simmer until needed.

Slice the steak and place on a bed of jus and Chanterelles. Garnish with tarragon leaves and serve.

horn of plenty

(*Craterellus cornucopioides*)

Trompette de Mort ('trumpet of death') is not the most encouraging of names for this beautiful and tasty fungus, but it does capture what it looks like rather well. The more squeamish could always try telling themselves it used to be called Trompette d'Amour ('love trumpet'), until some pessimist wrote the name down incorrectly. The Latin name, *Craterellus cornucopioides* is rather jollier – the second part means 'Horn of Plenty'.

The Horn of Plenty is related to the Chanterelle, and it shares something of that mushroom's apricoty aroma. It grows among the leaf litter of deciduous woodland during the autumn, where it is almost invisible, despite being highly gregarious. Depending on the weather conditions, its colour varies from black to brown to a beautiful mottled blueish grey.

Trompettes have a delicate, woody flavour, and a somewhat chewy or rubbery texture. They go extremely well with fish, especially turbot. The Horn of Plenty has been described as the Poor Man's Truffle, and used in its place in pasta dishes, but the reference is more a matter of appearance than of taste. In any case, these things are relative.

Dried Trompettes go for up to £25 a kilo.

'. . . this beautiful and tasty fungus'

Foie Gras with Horn of Plenty and Baby Leeks

Serves 4

Horn of Plenty have a delicious meaty flavour, and they go particularly well with foie gras. Particularly when sitting in a rich sauce made from beef stock, crème fraîche and dill.

This recipe is not ideal for those on a budget or a diet, but it is very, very nice.

250ml/9fl oz red wine jus (see recipe for Winter Chanterelles with Roast Beef and Red Wine Jus on page 56.)
150ml/1/4 pint crème fraîche
Salt and freshly milled black pepper
About 450g/1lb fresh foie gras
200g/7oz Horn of Plenty, cleaned and teased into strips
2 tsp chopped dill
8 baby leeks, boiled in water for 10 minutes, then cut in half lengthways

Bring the red wine jus to the boil. Add the crème fraîche, bring back to the boil and reduce by half. Season with salt and pepper.

Cut the foie gras into slices about 1cm/1/2in thick, season with a little salt and pepper and dry-fry in a nonstick pan on a medium heat until slightly browned on each side. The foie gras will give off lots of tasty fat. This is perfect for frying the Horn of Plenty.

Fry the Horn of Plenty for about 5 minutes, agitating frequently.

Add the dill to the crème fraîche jus sauce and pour on to 4 warmed plates.

Place the foie gras in the middle and arrange the Horn of Plenty and baby leeks around it.

Cod Baked with Horn of Plenty and Wild Garlic Leaves

Serves 4

Because of the parlous state of the cod fishing industry, we include this recipe in its present form with a certain reluctance. But if cod have to die, they might as well go out with a fanfare, this time of Trompettes de Mort.

4 firm cod fillets, about 2.5cm/1in thick
Juice of a lemon
1 sprig of flat-leaf parsley, finely chopped
1 sprig of mint, finely chopped
2–3 tbs olive oil
Salt and freshly ground black pepper
50g/2oz leeks, cut into fine strips
2 cloves garlic, finely chopped
50g/2oz unsalted butter
200g/7oz Horn of Plenty, cleaned and torn into strips
A handful of wild garlic leaves (or chives), roughly chopped
125ml/5fl oz crème fraîche

Wash and pat dry the cod fillets, then put aside.

Heat a nonstick baking tray in the oven for 10 minutes at 230°C/450°F/gas 8.

Mix together the lemon juice, parsley, mint, olive oil and a little salt and pepper.

Take the tray out of the oven, place the cod on it and spoon the mix liberally on top. Immediately replace in the oven. The cod will take 10 minutes at the most, during which time you should baste it once or twice. If it is milky white all the way through at the thickest point and the flakes separate easily, it is cooked.

While the cod is cooking fry up the leeks and chopped garlic with the butter in a pan. After a few minutes, add the Horn of Plenty and cook pretty fiercely to reduce the mushroom juices, stirring frequently as you go.

Lastly, add the wild garlic leaves, crème fraîche and salt and pepper. Take the cod fillets out of the oven and lay them on top of the mushroom sauce on warm plates.

Horn of Plenty and Gruyère Tart with Spinach

Serves 4

There is something particularly appealing in this recipe about the way the Horn of Plenty embed themselves in their cheesy under-layer. This tart goes very well with a sharply dressed salad.

Pie shell

125g/5oz plain flour
1/2 tsp salt
8 tbs unsalted butter
1 tbs grated Parmesan
1 tsp sugar
1 egg yolk
2–3 tbs cold water

The filling

1 red onion, finely sliced
50ml/2fl oz olive oil
200g/7oz Horn of Plenty, brushed clean and torn into strips
5 medium eggs
150ml/1/4 pint double cream
250g/9oz spinach, blanched in boiling water for a minute, then cooled and finely shredded
1/2 tsp salt
125g/5oz grated Gruyère cheese
A little ground black pepper

Combine the flour, salt, butter, Parmesan and sugar in a food processor for 10 seconds or so, until the mix looks crumbly. Then rub in the egg and water with a wooden spoon.

Flatten the mixture, wrap in cling film and refrigerate for an hour. Roll out the pastry and mould to the bottom of a 20–25cm/8–10in pie dish. Prick the bottom, then line the pastry with baking paper and weight down with dried pulses or rice.

Bake for 15 minutes at 180°C/350°F/gas 4. Remove the paper and bake for a further 15 minutes at 200°C/400°F/gas 6. Then set aside until the filling is ready.

Gently fry the red onion in the olive oil until soft, then add the Horn of Plenty and fry on for another 5 to 8 minutes.

Mix together the rest of the ingredients thoroughly.

Place the pie shell in the oven at 200°C/400°F/gas 6 for 10 minutes to warm it up.

Pour in the cheese mix, then spoon the mushrooms and red onion on top and stir in slightly.

Bake in the oven at 180°C/350°F/gas 4 for 30 to 40 minutes.

Lamb with Horn of Plenty, Couscous and Roast Peppers

Serves 4

The Horn of Plenty is an incredibly adaptable mushroom, which will go with practically anything. This Middle Eastern-inspired dish is a wonderful case in point, with a meaty lamb stock that the Trompettes love to soak up.

In Europe, fresh Horn of Plenty are available somewhere for rather more than half the year. Their season begins in the east in late summer, and finishes off in Portugal, where they grow into the early spring.

500g/1lb lamb leg fillet, cut into 2.5cm/1in chunks
Juice of a lemon
1 1/2 tsp ground cumin
2 cloves garlic, finely chopped
1 1/2 tsp ground paprika
Salt
2 red peppers
100ml/4fl oz olive oil
Black pepper
300g/10oz couscous
Enough lamb stock to cover the couscous, about 500–600ml/a generous 1 pint
250g/9oz Horn of Plenty, cleaned and torn into strips
50g/2oz sliced shallots
1 medium bunch of mint, leaves separated from stalks and roughly chopped
1 medium bunch of parsley, leaves separated from stalks and roughly chopped

Cover the lamb chunks in lemon juice, cumin, garlic, paprika and half a teaspoon of salt, and leave to marinate.

Roast the peppers in a very hot oven for 20 minutes. Shove in a plastic bag until cool, then peel the skin off, de-seed and slice into a julienne (i.e. very finely).

Take half the olive oil, a pinch of salt and half a teaspoon of ground pepper and mix thoroughly with the dry couscous. Heat the lamb stock to boiling point and pour over the couscous. Let it sit for 10 minutes, then gently separate the grains by lifting them then dropping them back down with a wooden spoon.

Fry the Horn of Plenty in half the remaining olive oil with the shallots over a moderate heat until they are pretty well cooked, about 5–10 minutes.

Add the peppers, Horn of Plenty and chopped herbs to the couscous. Mix in thoroughly, then check for seasoning.

Use the remaining olive oil to fry the lamb in a thick-bottomed pan over a moderate flame for 5–10 minutes, until nicely browned.

Heat up the couscous in the microwave, then serve with the lamb placed on top. If there is any couscous left over, it will be equally nice cold as a salad.

'. . . an incredibly adaptable mushroom, which will go with practically anything.'

blewits/ pied bleu

(*Lepista nuda*)

When it is young, the Wood Blewit (*Lepista nuda*) is among the most hauntingly beautiful of all mushrooms. As it ages, however, its intense almost iridescent lilac colour fades somewhat, particularly on the cap. What it retains throughout its life is a strong and distinctive perfume, a feature it shares with its relative the Field Blewit (*Lepista saeva*), also known as the Blue-leg.

Since blue in nature is usually a warning signal, it comes as a surprise to many that Wood Blewits are eminently edible, particularly when the young ones look so like dangerous sweets from a fairy tale. In fact, this suspicion is not entirely groundless: Blewits must never be eaten raw. Even when cooking has broken down the toxins they contain, a significant number of people remain mildly allergic to them.

Given this fact, it is quite a tribute to the Blewits' other qualities that even the mushroom-shy English have a long tradition of using them. In Berwick-upon-Tweed on the border with Scotland, there used to be an industry based on extracting dye from Blewits. The Field Blewit has been sold in markets in Derbyshire for centuries, and cultivated Blewits, grown outdoors on man-made mounds of compost, are even sometimes found in the supermarkets. And it's one of the relatively few mushrooms to have its own English name.

The Blewit's greatest virtue from the cook's perspective is probably its robustness, its ability to maintain its shape and texture through lengthy cooking processes. Blewits are excellent with game, and they start to appear after the first frosts.

'Since blue in nature is usually a warning signal, it comes as a surprise to many that Wood Blewits are eminently edible...'

Duck Breast with Blewits

Serves 4

For this aromatic, fruity dish you could equally well use Wood or Field Blewits, or their cultivated cousin the Pied Bleu. Blewits appear quite late in the season, occasionally well into December, so unless you are using cultivated ones, this could be your last fresh wild mushroom dish of the year.

4 duck breasts, skin scored at 1cm/½in intervals with a sharp knife
About 16 medium Blewits, brushed clean then blanched in boiling water for a few minutes
Salt and ground black pepper
Olive oil
½ tsp cinnamon
50ml/2fl oz red wine
1 orange, zested and squeezed
50ml/2fl oz chicken or duck stock
8 heads of pak choi, Chinese water spinach or regular spinach, chopped

Preheat the oven to 230°C/450°F/gas 8.

Season the duck breasts and Blewits with salt, pepper, olive oil and cinnamon, then cook in a baking tray in the oven for 20–30 minutes, depending on how well done you like your duck.

Remove the duck and the Blewits from the baking tray and skim off the fat from the juices with a tablespoon. Pour in the wine, orange juice and stock and place over direct heat. Let this liquor come to the boil, scraping the sides of the baking tray to incorporate the delicious flavours which will have accumulated during baking. Strain off the resultant liquid and reserve.

Cook the pak choi in boiling water for 3–5 minutes, then drain.

Slice the duck, then serve on top of the pak choi with the Blewits scattered on the plate. Pour wine sauce over each portion, and sprinkle with orange zest.

Chicken Pie with Blewits, Pumpkin and Sage

Serves 4

Blewits are highly perfumed mushrooms, which is part of their considerable appeal to chefs. But the extraordinary thing is that they are said to derive their scent from minute quantities of strychnine, aka rat poison. Not enough to do you any harm, you understand, but the thought does add a certain frisson to eating them.

1 red onion, finely sliced
200g/7oz Blewits, sliced
1 tbs butter
1½ tbs plain flour
150ml/¼pint chicken stock, heated up
250g/9oz pumpkin cut into 1cm/½in cubes, then simmered for 10 minutes or so in water until soft
400g/14oz chicken fillet, cut into small cubes
100ml/4fl oz double cream
A small bunch of sage leaves, finely chopped
Salt and ground black pepper
350g/¾lb puff pastry
1 egg, beaten with a drop or two of water

Preheat the oven to 230°C/450°F/gas 8.

Fry the red onion and the Blewits in the butter for about 10 minutes over a medium heat until soft. Add the flour, mix in, then slowly pour in the warm chicken stock, stirring vigorously to avoid lumps.

Remove from the heat and add the pumpkin, chicken, cream, sage and a little salt and pepper.

Transfer these ingredients to a pie dish, making sure that the liquid doesn't come above the solid ingredients, as this will make the pastry too damp.

Roll out the pastry on a floured surface and cut a long strip of pastry from the edge. Dampen the edge of the pie dish and press this strip all around the rim. Place the rolled pastry on top of the pie dish and trim flush around the sides. Squeeze down and flute the edge. Pierce a hole in the centre of the pastry to let out excess steam. Feel free to decorate it, maybe with a few mushroom shapes made of spare pastry.

Brush the pastry with beaten egg. Then place the pie in the oven on a baking sheet, (just in case it leaks,) for 10 minutes to brown it. Turn the heat down to 170°C/325°F/gas 3–4, cover loosely with foil and cook on for another hour or so. (Remember, when you turn the oven down it won't reduce its temperature immediately, so either leave the door ajar for a few minutes or move the pie down to the bottom of the oven.)

This dish is nice served with thick-cut chips and petits pois.

'...they are said to derive their scent from minute quantities of strychnine, aka rat poison.'

shaggy ink cap

(*Coprinus comatus*)

In 1943 the groundsman of a recently constructed soccer pitch in Stafford, England, arrived at work to find a most unusual problem. The entire two-acre site was white with mushrooms, each one resembling a miniature judge's wig. By the following day, his dilemma appeared to have spontaneously resolved itself. It was only on closer inspection that he realized at what price. For instead of a forest of mushrooms, the pitch was now covered with a sea of unsightly black goo. Somehow, the entire crop had turned to liquid overnight.

The rather beautiful technical term for what had happened is 'deliquescence'. It is by this process, a kind of auto-digestion, that the Shaggy Ink Cap (*Coprinus comatus*) paves the way for the release of its spores. And these spores are most inclined to germinate on land that has recently been disturbed – hence the Stafford experience.

As a result of the Shaggy Ink Cap's reproduction technique, the challenge for the mushroom hunter is to eat it before it starts to eat itself. It is the ephemeral nature of this mushroom as much as its delicate taste that makes it such a pleasure to consume.

The related Common Ink Cap (*Coprinus atramentarius*) is also edible, provided you do not drink alcohol with it, or indeed for several days afterwards. Should you do so, you will very much regret it, as you lurch between palpitations and nausea. This species contains an antabuse and has been successfully employed in the treatment of alcoholics.

Crispy Shaggy Ink Cap Fritters with Sour Cream

Serves 4

Because Shaggy Ink Caps are such fragile and transitory creatures, they respond best to rapid cooking techniques. This tempura-style recipe preserves their delicate flavour and texture extremely well. It also doesn't give them time even to think about liquefying.

The dip
200ml/7fl oz sour cream
1 tsp honey
1 tsp grain mustard
A squirt of lemon juice
1 tsp chopped dill
A pinch of salt and ground pepper

The fritters
1 egg yolk
250ml/9fl oz iced water
100g/4oz plain flour
Vegetable oil for deep-frying
12 small to medium Shaggy Ink Caps, brushed clean, stalks trimmed off flush with the cap
8 baby courgettes, blanched in boiling water for a couple of minutes

Mix together the ingredients for the dip and reserve.

Heat up the vegetable oil. You want it about 5–8cm/2–3in deep in a medium-sized pan.

Lightly stir together the ingredients for the batter, namely the egg, water and flour. Then flick a bit of batter into the oil. If it browns in a minute, the oil is ready.

Dip the Ink Caps into the batter, then fry in the oil until lightly browned. Then do the same with the courgettes. If the pan isn't big enough, fry the mushrooms and courgettes in two or three batches, keeping the others warm in a low oven.

Serve immediately. Dip in the sauce and place in the mouth.

Ink Caps with Rice Noodles in Broth

Serves 4

The Japanese aesthetic sensibility is particularly attuned to things that are fleeting – think of haikus, or the national obsession with cherry blossom. Capturing the precise moment is what matters. There is therefore something very appropriate about Japanese dishes which feature this most transitory of mushrooms. This example is clear, clean, aromatic and tasty.

Just over a litre/about 2 pints rich chicken stock
8 to 12 Shaggy Ink Caps, torn into strips
1 tbs sesame oil
2 tbs ginger juice, obtained by grating or blending a large knob of ginger, then passing the juice through a fine sieve or squeezing through a muslin
2 tbs tamari (much tastier than soy sauce)
1 tbs mirin
4 spring onions, sliced thinly
A pile of freshly cooked rice noodles, coated in a little sesame oil
1 tsp toasted sesame seeds, obtained by dry-frying in a pan for a couple of minutes over medium heat, stirring frequently

Heat up the chicken stock in a large pan.

Fry the Shaggy Ink Caps in the sesame oil over a medium flame until soft.

Pour the ginger juice into the stock, add the Ink Caps, tamari, mirin and spring onions, and simmer for a few minutes. Portion this broth into bowls of rice noodles, sprinkle with sesame seeds and serve.

parasol mushrooms

(*Macrolepiota procera*)

A mature, fully opened Parasol Mushroom is one of the most extraordinary sights in nature. It just looks too big to be real. The cap can be the size of an LP record, 30cm/1ft across, mounted on an implausibly flimsy-seeming stalk. It is quite obvious how this mushroom gets its common name.

Parasol Mushrooms grow in grassy places any time between July and November, often in fairy ring formation. Look for a pronounced brown 'nipple' in the centre of the caps, together with a smattering of snake-like scales radiating outwards. You should also be able to move the ring on the stalk up and down freely. Younger specimens, not fully opened, look rather like drumsticks.

This is one of the mushrooms that has recently had its name changed (to *Macrolepiota procera*). You will find it in older books under *Lepiota*, but this name now covers its smaller (and often deadly) cousins.

Parasols have a delicate flavour and texture which make for sublime tempura. The best and tenderest specimens are those that are just opening, although these lack some of the drama of the finished article. One cap should be sufficient for each person eating. The stem is too tough to be worth bothering with. If you manage to find any Parasol Mushrooms, enjoy them while you can. They are too hit-and-miss to be commercially traded. As a consequence, you will be eating something that money just can't buy.

Parasol Mushrooms with Corn Fritters

Serves 4

Parasol Mushrooms and sweet corn come into season at roughly the same time. The mushrooms are free, sweet corn is dirt cheap, and the rest of the ingredients aren't exactly expensive. So this is a luxury available to anyone, provided they have access to *Macrolepiota procera*. The fritters are delicious dipped in a yoghurt dressing mixed with lemon juice and tarragon.

4–6 Parasol Mushroom caps
50g/2oz butter
2 corn-on-the-cobs
100g/4oz plain flour
A pinch of salt and a grind or two of black pepper
1 egg
150ml/1/4 pint milk
1 tbs chopped parsley
Vegetable oil for frying

Clean and roughly chop the Parasol caps, then fry them in the butter until soft. Boil the corn-on-the-cob for 5 minutes, cool under running cold water, then shave the kernels from the cobs with a sharp knife, and reserve. (Mind your fingers.)

Sieve the flour, salt and pepper into a bowl, make a well in the centre and break in the egg. Add half of the milk and whisk until smooth. Slowly add the rest of the milk, whisking continually to retain the smoothness. Add the rest of the ingredients to the batter (but not the juices from the mushrooms, as this will make the batter too thin).

Heat up the vegetable oil in a pan until a shimmering haze appears above it. Now you are ready to fry.

Using a wet tablespoon, gather up mounds of batter and carefully lower each golfball-sized fritter into the oil. When nicely browned, remove with a slotted spoon, place on kitchen paper and keep warm until all are ready.

Parasol Mushrooms, English Muffins and Mature Cheddar

Serves 4

As the evenings start to close in and the rains come down, console yourself by gorging on this cosy snack.

The muffins

250ml/½ pint milk
1 tsp honey
1½ tsp instant yeast
450g/1lb flour
2 tsp salt
2 tbs butter
cornmeal

The topping

Butter
The caps of 4 large Parasol Mushrooms
Salt and freshly milled black pepper
About 150g/5oz grated mature Cheddar cheese

To make the muffins, begin by heating the milk to about 65°C/150°F, so you can just put your finger in without it hurting too much. Stir in the honey and the yeast.

Sift the flour and the salt into a large bowl and use your fingers to flake in the butter. Make a hole in the middle, pour in the milk and mix together to form a dough. If the dough seems too sticky, add a bit more flour. If it seems too dry, add a drop of water.

Knead the dough on a clean, flat surface. The ball should become very smooth and elastic.

Grease a bowl and place the ball of dough within. Cover with a plastic bag and leave in a warm place until the dough has doubled in size. This should take about 45 minutes.

Roll out the expanded dough on a lightly floured surface until it is about 1 cm/½in thick. Then shape your muffins with a small round cutter.

Flour a baking tray, place the muffins on it and cover. Let rise for another 45 minutes.

Sprinkle the muffins with cornmeal, then cook on top of the stove in a thick-bottomed frying pan or griddle, keeping the heat low, for 5 to 7 minutes on each side.

You can store these in an airtight container for a few days before using them.

To make the topping, first melt a large knob of butter in the frying pan. Add as many Parasol Mushrooms as will fit in the pan (i.e. about one), season with salt and pepper, and fry over medium heat for about 10 minutes per panful. Make sure to turn the mushroom caps as they fry.

Toast the muffins, then cut them in half. Pile with them with grated cheese and butter then grill.

You can either eat the Parasols sandwiched between two slices of muffin, or if this proves too unwieldy, you can resort to a knife and fork.

saffron milk caps

(*Lactarius deliciosus*)

The citizens of Pompeii were major fans of the Saffron Milk Cap, if a surviving fresco, which depicts it lovingly, is anything to go by. They may even have called it *Lactarius deliciosus*, the 'official' Latin name that it goes by today. Undoubtedly, they would have relished its firm, crunchy texture.

In their prime, Saffron Milk Caps are gently funnel-shaped, and somewhere between pink and orange. Older or bruised specimens stain a rather forbidding green. They grow in association with conifers, and in particular with the spruce tree. We have a contact in Scotland who reckons what they like most are spruce plantations between ten and fifteen years old.

All Milk Caps have a habit of secreting a milky substance when broken, hence the name. The Saffron Milk Cap, though, oozes a milk that is distinctly orange, unlike its relatives, whose milk tends to be, well, milky. When you break a piece off one, it audibly snaps, provided it isn't too far gone.

Insects love Saffron Milk Caps. So do the Poles, who take full advantage of how well they pickle. In the UK, this mushroom is imported from Australia out of season. Its texture makes it very good in stir-fries.

Grilled Sea Bass with Saffron Milk Caps

Serves 4

In the autumn of the year 2000, Nick caught quite a few Sea Bass in his inflatable kayak as they came into the shallows in search of bait-fish. Simultaneously, there was a glut of Saffron Milk Caps on land. This recipe was the inevitable result. You could almost say it was meant to be.

6 tomatoes
75ml/3fl oz olive oil
2 cloves garlic, finely chopped
2 shallots, finely chopped
1 small head of fennel, finely chopped
1 tbs capers, finely chopped
8 strips of anchovy in oil, finely chopped
A sprig of thyme, leaves picked off stalk
150ml/1/4 pint white wine
salt and pepper
8 medium Saffron Milk Caps, stalks trimmed flush to the cap
1 tsp paprika
8 small sea bass fillets, skin on and scaled, seasoned with rough sea salt and ground black pepper

Score the skin of the tomatoes with a sharp knife and dunk them in boiling water for 10 seconds. Then plunge them into iced water. The skin should peel off easily. Cut the skinned tomatoes in half, squeeze out the seeds and finely chop the flesh. Reserve.

Heat up a third of the olive oil in a nonstick pan, then add the garlic, shallots and fennel. Cook over a moderate heat for 10 minutes or so, until soft.

Add the capers, anchovies, thyme, tomatoes, white wine, pepper and a little salt (depending on how salty the capers and anchovies are).

Continue to cook on a moderate heat until the wine has reduced by half, stirring frequently. Pour off into a bowl.

Season the Milk Caps with a little salt, pepper and paprika and in the same pan fry them in half the remaining olive oil over a moderate heat until they are cooked through and slightly browned. Reserve.

Wipe the same pan clean, add the remaining olive oil and heat, place the fillets in, skin side down, and fry on a moderate heat for a couple of minutes on each side.

Serve the fillets on top of the sauce, skin side up, with the Milk Caps as garnish.

Saffron Milk Caps, Green Tagliatelle and Porcini Cream

Serves 4

When you collect Saffron Milk Caps, you can minimize bruising by making sure you place them in your basket gills upwards and only two deep. As you pick them, they will probably stain your hands orange. If you eat enough of them, they may also stain something else orange (clue: think beetroot or asparagus). Don't be alarmed if this happens. As Saffron Milk Caps tend to coincide with Ceps or Porcini (*Boletus edulis*) in both where and when they grow, it seems only natural to cook them together . . .

2 medium-sized Ceps, cleaned and finely chopped
4 tbs olive oil
150ml/6fl oz double cream
Half a leek, cleaned and cut into a julienne (thin strips)
2 cloves garlic, finely chopped
3 shallots, finely chopped
8 Saffron Milk Caps, cleaned and sliced
500g/1lb dry green tagliatelle
1 tsp chopped thyme
1 tsp chopped parsley
1 tsp chopped tarragon
75g/3oz grated Parmesan
Salt and freshly ground black pepper

Fry the Ceps (Porcini) in 2 tablespoons of the olive oil for 5 minutes or so over a medium flame. Pour in the cream and briefly bring to the boil. Place in a food processor and blend. Reserve.

Fry the leek, garlic and shallots in the rest of the olive oil for 5 minutes, stirring frequently, until soft. Then add the Milk Caps and continue to cook for another 5 minutes, until they have released their juices.

While you are preparing the sauce, the pasta can be cooked according to the instructions on the packet.

Add the Porcini cream to the Milk Cap mixture along with the thyme, parsley, tarragon, Parmesan and a little salt and pepper, and bring to a simmer for a minute. Combine with the pasta and serve.

hedgehog mushrooms

(*Hydnum repandum*)

It probably says a lot about our respective nations that the English name for *Hydnum repandum* is based on what the mushroom looks like underneath (from the viewpoint, say, of someone inspecting it nervously), while the French, who are a foraging people, describe it according to what it looks like from above. 'Pied de Mouton' means 'Sheep's Foot', and that is just what this mushroom is reminiscent of, with its white, irregularly lobed cap.

We debated putting the Hedgehog in the Oddballs chapter, as, unusually, it produces its spores on spines rather than gills. But as it makes excellent and fairly orthodox eating, and as it is quite important in the food trade, we've deemed it weighty enough to make the heavyweight division.

Young Hedgehogs are lovely and tender. Older ones may need blanching to take away a slight bitterness. You will also probably want to remove the spines, which will otherwise break off and get everywhere, perhaps spoiling the appearance of whatever you are making. The Pied de Mouton keeps its texture particularly well when it is cooked.

In Britain, the Hedgehog Mushroom grows in various kinds of woodland from September to early December, but in Portugal it carries on right through the winter. An export industry has grown up as a result.

Hedgehogs Swimming with Scallops and Water Spinach

Serves 4

This is a Japanese-inspired soup, which uses Hedgehog Mushrooms where it could equally well use Shiitake. An Oriental cook would make this a filigree of finely shredded ingredients; westerners might opt for the chunkier, robust offering shown here.

1 litre /1¾ pints light dashi (Japanese fish stock) – see below
1 tbs juice from a knob of fresh ginger
About 12 leaves of pak choi, spinach or other green leafy veg of your choice
2 tbs tamari (light soy sauce)
2 tbs mirin (Japanese sherry)
100g/4oz Hedgehog Mushrooms, very finely sliced
4 large scallops

To make the *dashi* the traditional way, add a 15cm/6in piece of *kombu* (dried kelp) to 1 litre/1¾ pints of water, bring to the boil and then take out the kombu. Remove from the heat and immediately add 15g/½oz dried bonito flakes. Allow them to settle, then gently strain the liquid through a fine sieve or muslin. If, on the other hand, you feel that life is too short, just buy some instant *dashi* in an Oriental supermarket and follow the instructions. Much of Japan will be on your side.

To obtain the ginger juice, blend a knob of ginger in the food processor. Squeeze the juice from the resulting pulp and pass it through a sieve. Briefly blanch the pak choi leaves in boiling water, then plunge into cold water and drain.

Place the dashi in a pan with the tamari, mirin and ginger juice and simmer for 2–3 minutes. Stir in the Hedgehogs and simmer for 5 more minutes. Slice in the scallops, add the leaves and serve immediately.

As you never know quite what you will come home with after a day's foraging, we close this chapter with a few handy, all-purpose recipes which it should prove useful to have at your disposal.

Wild Mushroom Lasagne

Serves 4

This delicious lasagne will work with almost any mushroom you care to mention. Fresh Ceps would be our first choice, but dried ones can impart a fantastic flavour. Nick has also made this dish with fresh Morels, which was a delight. You will need a rectangular lasagne dish about 30 x 25cm/12 x 10in, or one very slightly larger.

The béchamel

750ml/1 1/3 pints milk
175ml/7fl oz double cream
50g/2oz butter
75g/3oz plain flour
1/2 tsp salt and a little pepper
1/2 tsp ground nutmeg
100g/4oz grated Parmesan

The spinach layer

300g/10oz spinach, washed, de-stalked, then dunked in boiling water for a minute until wilted. Take out, refresh in cold water, and then finely chop
250g/9oz Ricotta
1/2 tsp salt
A generous grind of black pepper

The mushroom sauce

2 cloves garlic, chopped
150g/6oz shallots, finely diced
75ml/3fl oz olive oil
300g/10oz mixed mushrooms, sliced
A couple of sprigs of thyme, chopped
A handful of chopped flat-leaf parsley
1/2 tsp salt
A generous amount of ground black pepper
200ml/7fl oz crème fraîche

The lasagne

Either use fresh or dried lasagne sheets. Make sure you have enough for 3 layers. About 300g/10oz if you are using fresh; one 250g/9oz packet if you are using dried.

First make the béchamel. Heat the milk and cream to a simmer.

Meanwhile, melt the butter in a thick-bottomed saucepan, add the flour and cook the resulting sandy paste over a low flame for a couple of minutes, taking care not to burn.

Keeping the heat low, slowly pour in a little milk. Whisk until thick and smooth. Then add a little more milk, again whisking until smooth, and so on. Continue until all the milk has been added.

Season with the salt, pepper and nutmeg and sprinkle in half of the Parmesan. Whisk until smooth. Cook slowly for 10 minutes, stirring continuously to prevent the sauce from burning on the base of the pan. Reserve, with a little greaseproof paper on top to prevent a skin from forming.

Now mix together the ingredients for the spinach layer in a bowl, and reserve.

To make the mushroom sauce, fry the shallots and garlic in the olive oil for a few minutes over a moderate heat, then add the mushrooms and cook on for 10–15 minutes, until they are cooked.

Add the rest of the ingredients and, stirring frequently, reduce the sauce a little over a low to moderate heat.

You shouldn't need to precook the lasagne sheets, but if you are using dried ones, check the packet before use. Some brands need precooking.

To put the dish together, begin by spreading a little béchamel on the base of the dish, then place a sheet of lasagne on top. On this evenly spread the Ricotta mix. Now add another layer of lasagne, and pour the mixed mushroom sauce on top. Next comes the last layer of lasagne and finally the béchamel sauce. Sprinkle this with the last of the Parmesan.

Cook in the oven at 200°C/400°F/gas 6 for 45 minutes. If the lasagne needs a little more browning after cooking, it can be finished off under the grill.

Polenta with Wild Mushrooms

Serves 4

This recipe will liven up plain old polenta no end. What's more, it can be made with practically any type of mushroom, from cultivated Button Mushrooms masquerading as Field Mushrooms to Ceps. Experience tells us that it goes particularly well with cubed pancetta and roast tomatoes sprinkled with Parmesan.

1.25 litres/2 pints chicken or vegetable stock
300g/10oz finely diced Ceps, or other mushrooms of your choice
2 cloves garlic, chopped
2 tbs finely chopped shallots
100ml/4fl oz olive oil
Juice of a lemon
250g/9oz quick-cook polenta
A sprig or two of thyme, finely chopped
1 tbs finely chopped flat-leaf parsley
50g/2oz grated Parmesan
Salt and freshly ground black pepper

Heat up the chicken or vegetable stock.

Gently fry the mushrooms, garlic and shallots in half the olive oil until soft, then add the lemon juice.

Pour the polenta into the simmering stock, whisking continually for a few minutes until the mixture has thickened considerably.

Add the mushrooms, thyme, flat-leaf parsley, Parmesan and a generous amount of salt and black pepper.

You can eat this dish as it is. Or you can line an earthenware dish with greaseproof paper and pour the polenta mix in. Let it set for an hour or two, then slice thickly. Fry in a griddle pan until browned in the remaining olive oil.

Wild Mushroom Pancakes

Serves 4

It is difficult to tell in advance quite how much time and creative energy you and your fellow-foragers will have at your disposal after a day's hunting.With the right sort of inspiration, this pancake can become a veritable work of art, with a different coloured mushroom used for each layer. Reduced to its bare essentials, as in the photo overleaf (i.e. a randomly stuffed mushroom pancake), it is still thoroughly good.

The pancakes
2 eggs
250ml/½ pint milk
2 tbs melted butter
2 cups plain flour
1 tsp baking powder
½ tsp salt
Butter

The mushroom filling
300g/10oz wild mushrooms, ideally two or three different kinds
300g/10oz pumpkin
50ml/2fl oz crème fraîche
Salt and pepper
100g/4oz unsalted butter
2 cloves garlic, finely chopped
4 shallots, finely diced
100ml/4fl oz white wine
1 sprig of parsley, chopped
150ml/5fl oz additional crème fraîche
Salt and pepper

Clean and roughly chop the mushrooms and reserve, keeping each species separate.

Peel, de-seed and chop the pumpkin and boil in water for 20 minutes. Drain and then blend with the crème fraîche and the seasonings.

To make the pancakes, begin by beating together the eggs, milk and melted butter.

Mix all the dry ingredients together, and slowly beat into the egg and milk mixture.

Make each pancake by heating a teaspoon of butter in a nonstick frying pan, then ladling in a small amount of the batter. Cook fiercely, and toss to ensure a light browning on the other side. Store in a warm oven, piled up and covered with silver foil.

Divide the rest of the ingredients (except the pumpkin mixture) according to how many varieties of mushroom you have at your disposal. Using one species for each layer will be good for the aesthetics.

Fry the mushrooms in batches in the butter with the garlic and shallots. Once they have released their juices, add the wine and reduce by half, then stir in the parsley and crème fraîche and a little salt and pepper.

Heat the pumpkin mix, either blend with the mushrooms and slop onto the pancakes unceremoniously, or pour a pool onto a warm serving plate and place a mushroom-stuffed pancake alongside. Or use your artistic skills to build a pancake 'millefeuille'. Serve immediately.

chapter 4

the taming of the shroom

In early January, Kiruna in Swedish Lapland seems a distinctly unpromising place for mushrooms. In the depths of the month-long Arctic night, where the temperature can plunge to minus 35°C and below, the huge mountain which usually dominates the town is reduced to a mere suggestion. Yet deep within that mountain, hollowed out from years of iron-mining, a fervour of fungal activity continues throughout the year. All because of the vision of one fungophile Swede, who realized the cool, damp, abandoned mineshafts would be perfect for growing Shiitake.

The Shiitake (*Lentinula edodes*) has come a long way since it became arguably the first mushroom in the world to be cultivated by human beings. Over a thousand years ago, residents of the Chinese province of Qingyuan began to experiment with grafting wood from Shiitake-producing logs into old living trees. Around the year 1200, the technique was refined by one Master Wu San Kwung from the village of Longyan, thereafter known to the locals as the God of the Mushroom. His method is described in depth in Wang Cheng's classic agriculture book of 1313.

The traditional method of growing Shiitakes begins with the inoculation of sawn logs of oak (or in Japan, the *shii* tree) with mycelium harvested from infected wood. The hungry farmer then has to wait for between two and four years for the mycelium to spread sufficiently for the first fragrant, woody mushrooms to appear. With luck, a log will then produce Shiitake for up to six years. In contrast, modern techniques involving sterilized man-made logs have reduced the gap from inoculation to first fruiting to seven weeks, and the entire life-cycle to around four months. It is this revolution that has allowed the Shiitake to become the second most important mushroom crop in the world, with over two million tons grown annually.

Watercress and Shiitake Salad with Tofu

Serves 4

This is fresh, invigorating and, according to Nick's Chinese wife, good for longevity and good for the skin. It is definitely good for the stomach.

1 tbs soy sauce
A block of fresh tofu, about 250g/9oz, cut into thick slices
Vegetable oil for shallow-frying, seasoned with 2 tbs sesame oil
250g/9oz fresh Shiitake caps, thinly sliced
2 tbs sesame oil
1 tsp finely chopped red chilli
1 tsp finely chopped ginger
1 tsp finely chopped garlic
1 sprig of fresh coriander, roughly chopped
1 tbs fish sauce
A bunch of watercress, washed and cut into manageable pieces

For the watercress dressing
1 tsp toasted sesame seeds
1 tbs sesame oil
1 tbs honey
2 tbs soy sauce
Juice of 1 lime

Splash a little soy sauce on the tofu and shallow-fry in the vegetable and sesame oils mixed together. Cook until slightly golden on each side.

Stir-fry the mushrooms in 2 tablespoons sesame oil with the chilli, ginger and garlic, the coriander and the fish sauce, until the mushrooms have released their juices and these have reduced by half.

Mix together the ingredients for the dressing and toss with the watercress in a large bowl.

Place the mushrooms and the tofu on top of the watercress, either as individual portions or in a large bowl.

Shiitake Stir-Fry with Plum Sauce

Serves 4

Home-made plum sauce is a million miles away from the shop-bought variety and it brings a mellow and fruity spiciness to this stir-fry. This recipe belongs to Nick's friend Camilla, who has an ancient plum orchard in her back garden. Having been untouched by pesticides and fertilizers, and grazed only by sheep, this unkempt patch now sprouts an amazing variety of wild mushrooms. These include Horse Mushrooms, Chicken of the Woods and Morels – but not, unfortunately, Shiitake. Happily, they are now readily available in supermarkets. You can also grow your own – see page 157.

This recipe should produce four to six jars of plum sauce, which will last you at least a year. The fact that it is wonderful with duck will help you get through it, as will its excellent potential as a gift. This stir-fry is delicious on top of a pile of thick ribbon rice noodles.

The sauce

4 tbs sunflower oil
2 large onions, finely chopped
2 large knobs of fresh ginger, finely grated or chopped
3 cloves garlic, chopped
1 medium-strength red chilli, finely chopped
2.5kg/5 1/2lb dark plums or damsons, stoned
1 litre/1 3/4 pints vinegar. (Camilla uses half white and half red wine.)
300ml/1/2 pint water
250g/9oz soft brown sugar
200ml/7fl oz dark soy sauce
1/2 tsp cinnamon
1 spice bag (4 star anise and 1 tsp Sichuan pepper, all crushed and tied up in muslin)
Salt to taste

The stir-fry

300g/10oz fresh Shiitake (stalks trimmed off), sliced
2 cloves garlic, finely chopped
A knob of ginger, finely chopped
100g/4oz sliced shallots
1 tbs sesame oil
2 tbs vegetable oil
200g/7oz baby corn
300g/10oz pak choi, spinach or other green leafy vegetable of your choice
1 tsp sesame seeds, toasted by tossing in a pan over medium heat for a couple of minutes until slightly coloured

To make the sauce, avail yourself of a large thick-bottomed saucepan and pour in the oil. Cook the onions, ginger, garlic and chilli in it until nice and soft, then add the plums and the vinegar and water. Bring slowly to the boil, then simmer until soft.

Push the ingredients through a sieve using a wooden spoon, then return the purée to the pan. Add the sugar, soy sauce and cinnamon and the spice bag. Add salt to taste.

Bring to the boil, invite a friend round, and get them to keep stirring the sauce over medium to low heat while you sterilize 6 clean jam jars in a low oven for half an hour or so.

When the sauce is ready, take out the spice bag and fill the jars immediately almost up to the top. Cover tightly, then leave them to settle in a cool place. After opening they must be stored in the fridge.

To make the stir-fry, begin by frying up the Shiitake, garlic, ginger and shallots in the sesame and vegetable oils until soft.

Meanwhile, blanch the baby corn in boiling water for 5 minutes, and the pak choi, roughly chopped, for 2. Place both vegetables in a sieve and cool under cold running water.

Add the sesame seeds, pak choi and baby corn to the pan, plus 4 generous tablespoons of plum sauce.

Heat thoroughly and serve on a bed of rice noodles.

Beef Stewed in Shiitake Broth

Serves 6

The intense flavour of dried Shiitake suits this soup well. It is very simple to make – all you need is a few hours to let it simmer away while the flavours mingle. If you don't eat beef, try making this with chicken or pork.

1kg/about 2lb beef brisket
1.25 litres/2 pints chicken stock, or enough to cover the beef
1 cinnamon stick
1 tsp medium-strength chopped red chilli
2 cloves garlic
1 tsp chopped ginger
3 star anise
10 large dried Shiitake
450g/1lb frozen petits pois
4 spring onions, chopped
Soy sauce

Place the meat and stock in a large pan with the cinnamon, chilli, garlic, ginger, star anise and the Shiitake. Simmer for 3 hours, skimming frequently to prevent the soup from becoming cloudy. Don't let the broth boil freely, as this would adversely affect its clarity and taste.

Remove the soup from the heat and strain, reserving the meat, Shiitake and stock. When the meat has cooled, cut it into thin slices or pull it into shreds, then thinly slice the mushrooms.

Return the stock to the heat in a clean pan and when it is simmering, add the beef, petits pois, Shiitake and spring onions. Season with soy sauce and serve with rice or noodles.

Poached Salmon with Shiitake and Black Bean Sauce

Serves 4

This is a wonderful, aromatic dish with that makes the most of your Shiitake.

200ml/7fl oz chicken stock
Juice of 2 limes
50ml/2fl oz rice wine or sherry
2 tbs soy sauce
2 tbs sesame oil
4 skinless salmon fillets, about 150g/6oz each
1 tbs finely chopped ginger
2 cloves garlic, finely chopped
4 spring onions, finely sliced
150g/6oz thinly sliced Shiitake
1 tbs vegetable oil
100g/4oz roughly chopped black beans (available in tins from Chinese grocers and large supermarkets)
2 tbs chopped coriander (optional)

Pour the chicken stock, lime juice, rice wine or sherry and half the soy sauce and sesame oil into a frying pan.

Heat the liquor up to a simmer, then place the salmon fillets gently in the pan.

Cook slowly, gently agitating the salmon. The liquor may not quite cover the fish, so turn the fillets over after 5 minutes and take the pan off the heat. The salmon will continue to cook in the tangy liquid.

Meanwhile, fry the ginger, garlic, spring onions and Shiitake quite fiercely in the remaining sesame and vegetable oils for 5 minutes, or until the mushrooms look cooked (when a Shiitake is ready, it feels a little slimy, but don't burn yourself testing for this).

Add the black beans and cook for a minute. Pour in the liquor from the pan and boil for a few minutes.

Pour the sauce over the salmon, garnish with the coriander if you like the stuff, and serve.

The Orientals may have been quick off the mark in mushroom cultivation in human terms, but they are mere novices compared with certain species of ants and termites, who have been at it for millions of years. The leaf-cutter ant of Central and South America, for example, does what it does with one purpose in mind: to provide compost for gardens of fungus buried deep within its nests. The fungus then breaks down the indigestible cellulose in the leaves, converting it into hyphae on which the ants are able to graze. This symbiosis has gone on for so long that neither ant nor fungus is now able to survive without the other.

There was something ant-like about the earliest successful European attempts to cultivate mushrooms, which again produced a subspecies no longer found in nature (*Agaricus bisporus*). The process began when the French king Louis XIV, entranced by the white Button Mushroom, ordered Olivier de Serres, his chief agronomist, to find a way to grow it all year round. Understandably keen to avoid disappointing the Sun King, de Serres moved heaven and (literally) earth in his attempts to oblige. Fortunately, he hit on a solution despite never quite knowing why it worked: if mature specimens were dug up and replanted in a spot which duplicated the conditions of their original sprouting, others would appear shortly thereafter.

It was not, however, until the end of the eighteenth century that something was noticed that really allowed the French to capitalize on their new discovery. In spite of what everyone had hitherto assumed, it turned out that White Mushrooms did not require light in order to grow. This meant they could be cultivated indoors, and Paris had perfect natural facilities in the form of the cool limestone caves that surround the city. Thus was the Champignon de Paris born, and with it the world's first mushroom millionaires.

Char-grilled Field Mushroom, Avocado and Chicken Salad

Serves 4

If you purchase 'Field Mushrooms' in the supermarket, they are likely to be specimens of *Agaricus bisporus* which have been allowed to mature to the point where their caps are fully opened. This salad, which shows them off to excellent effect, is best served warm.

4 large open-capped Field Mushrooms, seasoned with a little paprika, salt and pepper and brushed with olive oil
2 chicken breasts, sliced thickly and seasoned with lemon juice, salt, black pepper and 1 clove garlic, finely chopped
1 tbs olive oil
1 red onion, thinly sliced
50ml/2fl oz rich, home-made chicken stock
1 large ripe avocado, peeled and de-stoned, sliced thinly and dribbled with lemon juice to keep it green
12 cherry tomatoes, sliced
A sprig of tarragon, finely chopped
Juice of half a lemon

Grill the mushrooms on both sides until nicely browned, and reserve. (The barbecue would work well; alternatively, use a griddle pan.)

Fry the chicken in the olive oil in a hot pan until it starts to colour. Add the onion and continue to fry for 5 minutes until it is soft.

Pour in the chicken stock and reduce by half.

Dump the contents of the pan into a large bowl and throw in the avocado, cherry tomatoes, tarragon and lemon juice.

Place a mushroom cap on a warmed plate and arrange a portion of the salad mixture on top. Serve immediately.

Fish Pie with Smoked Haddock and Baby Button Mushrooms

Serves 4

700g/1½ lb potatoes
150ml/¼ pint milk
150ml/¼ pint cream
2 heaped tbs butter
150g/6oz finely chopped onion
150g/6oz finely chopped leek
250g/9oz whole baby Button Mushrooms, stalks trimmed flush with the cap
1 heaped tbs plain flour
Salt and ground black pepper
A couple of sprigs of thyme, chopped
2 bay leaves
1 tbs chopped parsley
300g/10oz undyed smoked haddock fillets, skinned and roughly chopped
150g/6oz mature Cheddar cheese, grated
1 tsp paprika

Peel and roughly slice the potatoes, then boil for 30 minutes or until slightly overcooked. Reserve.

Combine the milk and cream. Heat in a small pan to a gentle simmer. Melt 1 tablespoon of the butter in a large pan and fry the onion, leek and mushrooms over a medium heat for 15 minutes, or until the vegetables are nice and soft but not browned. Stir frequently as you cook.

Turn the heat down, then stir the flour in thoroughly with a wooden spoon until it has all been soaked up by the juices. Slowly pour in the heated cream and milk and beat with the spoon until all the lumps have disappeared.

Add salt, black pepper, thyme, bay, parsley, the smoked haddock fillets and half the Cheddar cheese. Simmer for 5 minutes or so, stirring frequently. Keep the heat down, or the sauce may burn. Add the remaining butter to the potato, season with salt and pepper, and mash.

Place the haddock mix into a baking dish, then evenly cover with the mashed potato. Crumble the Cheddar on top, then sprinkle with the paprika. Bake in an oven preheated to 220°C/425°F/gas 7 for 30 to 40 minutes until nicely browned on top This pie can easily be prepared in advance and stored in the fridge for up to two days before baking.

Today, global production of farmed mushrooms exceeds eight million tons per year, with around fifteen species cultivated on an industrial scale. Livesey Brothers' Oyster Mushroom farm in rural Leicestershire is typical of the high-tech world of the modern mushroom facility. Yet despite the sophisticated production process, the proprietors remain convinced that their work is as much art as science.

The first thing the visitor is shown is a huge and distinctly low-tech pile of steaming straw compost. This ferments away at up to 80°C until it has broken down sufficiently to take on the character of decaying wood, whereupon it is steam-treated to kill off any bacteria that might rival the soon-to-be added mycelium. It is then cooled and mixed with grains coated in mycelium taken from a 'mother culture' housed in a laboratory in France.

The inoculated compost or substrate is formed into bales, which are wrapped in plastic with slits cut into it to allow the exchange of gases. The bales sit in a warm, carbon dioxide-rich growing room for two to three weeks, by which time they are thoroughly riddled with Oyster Mushroom mycelium.

At this point, action must be taken to encourage the bales to fruit – something the lazy mycelium is reluctant to do if it can possibly help it. The ingenious solution to its reticence is to threaten it. At the appropriate juncture, the virtual season is changed suddenly and dramatically, by adjustments to the air-conditioning and climate control present in each of the farm's forty sheds. Humidity rises and the temperature plunges, shocking the mycelium into believing death is around the corner. This, it senses, may be its last chance to reproduce. So it fruits.

The clumps of Oyster Mushrooms that sprout through the slits in the plastic covering of each bale are reminiscent of nothing so much as outcrops of tropical coral. Three strains of the fungus are grown at Livesey Brothers: the standard velvety-grey variety, a yellow sort suggestive of

Dr Seuss trumpet keys, and an alarmingly vivid pink one. Whatever the vagaries of the English weather, at any one time there are twenty summers and twenty winters in progress within the farm's huge half-pipe sheds. Yet despite the elaborate lengths taken to create convincing artificial climates, the mushrooms somehow know what the weather is like outside. No one is quite sure how they know, but know they do. It shows in the rate at which they grow. Maybe they are sensitive to subtle changes in atmospheric pressure . . .

Chicken in Oyster Mushroom Sauce with Tarragon

Serves 4

A friend of Nick's lives close to a mushroom farm. Some of the spores must have escaped, because the last time he was there, he found the wooden frame of the greenhouse and an outdoor bench covered in Oyster mushroom. His host was astonished when Nick picked and ate them.

Duck would make a very good substitute for chicken if you were in the mood. This dish is delicious with sauté potatoes.

6 ripe tomatoes
2 tbs butter
4 chicken portions, seasoned with a little salt and ground pepper
100g/4oz finely chopped shallots
2 cloves garlic, finely chopped
250g/9oz Oyster Mushrooms, torn into thin strips
150ml/1/4 pint white wine
A handful of tarragon, finely chopped
Salt and pepper

Score the tomatoes with a sharp knife, dunk them into water on a rolling boil for 10 seconds, then plunge them into iced water. The skin should almost fall off, which you should encourage. Cut them in half, squeeze out the seeds and finely chop. Reserve.

Preheat the oven to 220°C/425°F/gas 7.

Melt half the butter in a frying pan and fry the chicken for 3 minutes on each side. Place in an oven dish and cook in the oven for 20 to 30 minutes, depending on how large the portions are. In the same pan as you fried the chicken, fry the shallots and garlic in the rest of the butter over a moderate flame until slightly browned. Add the mushrooms and continue to cook until they have released their juices and these have evaporated a little. Pour in the wine and tomatoes and reduce the liquid by half, stirring continuously, over moderate heat.

By this point the chicken should be ready. Take it out of the baking dish, skim off the fat with a spoon, then de-glaze with the sauce in the pan. Pour the sauce back into the pan. Add the tarragon and salt and pepper to taste and serve with the chicken.

Horse Mushrooms with Spinach and Pasta

Serves 4

Although the Horse Mushroom (*Agaricus arvensis*) isn't cultivated, we've elected to deal with it here alongside its close relative which is. And in fact, this recipe would work almost as well with Portabello or supermarket 'Field Mushrooms'. Choose whatever pasta you like. Rigatone holds the sauce well; mushroom ravioli would be very glamorous.

Recently, Nick drove past a field absolutely heaving with Horse Mushrooms. As he was on his way to a hotel where he wouldn't have access to cooking facilities, this put him in something of a dilemma. His solution was to restrict himself to a collecting a small bag full, which he donated to the chef on arrival. The latter was suspicious at first, perhaps because of the unfamiliar aniseed aroma. But later he caught up with Nick to sing a hymn of praise to the Horse Mushroom.

250g/9oz washed spinach
300g/10oz fresh young Horse Mushrooms, cut lengthways down the middle
50ml/2fl oz olive oil
100g/4oz shallots, finely chopped
100g/4oz leeks, julienned (cut into fine matchsticks)
2 cloves garlic, finely chopped
700g/1½ lb fresh pasta
1 tbs chopped flat-leaf parsley
½ tbs chopped tarragon
50g/2oz freshly grated Parmesan
250ml/½ pint double cream
Salt and freshly ground black pepper

Blanch the spinach in boiling water for a couple of minutes, then drain, cool and squeeze it out. Chop it up finely and reserve. Put on a large pan of boiling water for the pasta.

Fry the Horse Mushrooms quite fiercely in half the olive oil so they brown up nicely, then remove from the pan. Fry the shallots, leeks and garlic quite gently in the rest of the olive oil for 10 minutes or so, until soft. Then replace the mushrooms. Meanwhile, cook the pasta.

Add the herbs to the sauce, with the spinach, Parmesan and the double cream, and season to taste. Bring briefly to the boil, stirring continually. Take the pasta off the heat, drain it and mix in the sauce. Serve immediately.

Potato Croquettes with Roast Garlic and Field Mushrooms

Serves 4

These croquettes make a tasty starter – delicious with a dipping sauce of sour cream and dill with a dash of lemon juice. We have used Field Mushrooms, but feel free to experiment with other varieties, such as Oysters or Saffron Milk Caps.

200g/7oz Field Mushrooms
A generous 50g/2oz butter
250ml/9fl oz milk
1 blade of mace
1 bay leaf
1 small onion
50g/2oz flour, plus a little extra for coating the croquettes before breadcrumbing
150g/6oz potato, cooked and mashed
1 head garlic, roasted (see page 43)
Salt and freshly ground pepper
25g/1oz grated Parmesan
2 egg yolks
1 tbs chopped flat-leaf parsley
1 whole egg
Breadcrumbs for coating
Oil for deep-frying

Finely chop the mushrooms, then fry gently in a knob of the butter for 10 minutes, and leave to cool.

Heat up the milk with the mace, bay leaf and onion. Simmer for 10 minutes.

Melt the butter over medium heat in a thick-bottomed pan, then stir in 50g/2oz flour. Continue to cook for a further 3–5 minutes, stirring with a wooden spoon.

Slowly pour the warmed milk into the flour paste through a sieve, whisking the sauce frenetically at the same time. Take care not to burn yourself with the hot liquid.

After this process you should have a smooth, lump-free sauce. Cook on for a couple of minutes over very low heat, then take off the stove. Add the potato, the squeezed garlic, mushrooms, salt, pepper, Parmesan, egg yolks and flat-leaf parsley to the sauce, mix in well, then taste for seasoning and adjust if necessary. Transfer the mix to a shallow tray and place in the fridge to cool.
Make small cylindrical shapes out of the croquette mix ready for coating with breadcrumbs. (Alternatively, roll it into balls.)

Break the whole egg into a small bowl and lightly whisk with a little salt and pepper and a teaspoon of oil. Place a pile of breadcrumbs on greaseproof paper.

Lightly flour each croquette first, then dip in the beaten egg. Roll to and fro in the breadcrumbs until completely coated. Let the croquettes rest in the fridge for half an hour before cooking.

Heat up the oil for frying in a deep pan. Make sure there is enough to cover the croquettes. When the oil reaches 190°C/375°F (hot enough to brown a piece of bread in 20 seconds), it is ready for use.

Fry the croquettes in small batches until crisp and browned.

Rice Salad with Oyster Mushrooms and Pak Choi

Serves 4

Wild Oyster Mushrooms are a real trophy, and we have come across them at all times of the year. There are a few different varieties in this country. Angels' Wings, a pure white form, colonize dead wood in the Scottish Highlands. *Pleurotus cornucopiae*, in all its creaminess, may be found on dead oaks in Wales. However, the most common variety is the grey-to-brown Pleurotus ostreatus, which often grows on dead beech wood. You will need to check wild Oysters for maggots and brush off any dirt, but you should be all right with shop-bought ones.

If pak choi is unavailable, any other green leaf vegetable will do, such as spinach or shredded cabbage. This dish is delicious hot or cold. It is a terrific accompaniment to squid, which should be sliced, seasoned with lime juice and soy sauce, and grilled or fried.

350g/12oz basmati rice
4 heads of pak choi, leaves separated
1 tsp finely chopped red chilli
1 tsp finely chopped ginger
1 tsp finely chopped garlic
1 tbs sesame oil
2 tbs vegetable oil
300g/10oz Oyster Mushrooms, torn into strips
2 tsp sesame seeds, dry-fried for a couple of minutes until aromatic and slightly brown
A handful of fresh coriander, finely chopped
1 tbs fish sauce
2 tbs soy sauce
A small bunch of spring onions, finely sliced
Juice of a lime

Rinse the rice, then cook it slowly in a rice cooker or a saucepan with a tight-fitting lid for 15 to 20 minutes. To gauge the correct quantity of water to use, measure the depth of the rice with your finger, then cover it with enough water to add a third to its height. After you have taken it off the heat, leave the lid on for 5 minutes before serving.

Blanch the pak choi in boiling water for 3 minutes, then cool it under the tap.

Gently fry the chilli, ginger and garlic in a mixture of the sesame and vegetable oils for a couple of minutes. Add the Oyster Mushrooms.

Continue to cook, stirring frequently. The mushrooms tend to give off quite a lot of liquid. Cook until this has been released and evaporated.

Place the rice in a large bowl. Add the sesame seeds, coriander, fish sauce, soy sauce, pak choi, spring onions, lime juice and mushrooms. Mix the ingredients together thoroughly and serve.

Other important cultivated varieties include the Brown 'Crimini' strain of the common domesticated *Agaricus* mushroom, known in its open-capped adult form as the Portabello; the Blewit (see Chapter 3 on the 'Wild Heavyweights'), which can be grown outdoors in soil-covered mounds, and Japanese specialities like the long-stemmed Enoki, star of the delicate recipe which closes this chapter. Special mention should also be made of the chewy, resilient Paddy Straw Mushroom, *Volvariella volvacea*, so important in the Orient that it accounts for about 15 per cent of the world's total mushroom crop. For this alone it merits its recipe.

With ample economic incentive provided by the growing public appetite for exotic mushrooms, several new species are likely to be cultivated in the years to come. Meanwhile, at the margins of science, visionary mycologists are busily investigating ways in which cultivated fungi might form the basis for self-sustaining artificial environments, as man forges ever further into space.

Paddy Straw Mushroom and Prawn Stir-fry

Serves 4

One of the reasons Paddy Straw Mushrooms are so popular in the Far East is that the growing substrate they like best (straw from rice paddies) is predictably abundant in those parts. But the real key to their appeal lies in their extraordinary texture, particularly in the 'egg' stage at which they are usually eaten. Imagine a deliciously mushroomy eyeball, and you'll be part of the way there.

2 tbs peanut oil
150g/6oz peeled raw tiger prawns (tails left on), seasoned with a little soy sauce, 1/2 red chilli (chopped), 1/2 tsp chopped ginger and a chopped clove garlic
1 tsp sesame oil
1 clove garlic, crushed
1 red chilli, chopped
1 tsp chopped ginger
6 spring onions, chopped
1 tbs dry sherry
1/2 tsp sugar
A 400g/14oz can of Paddy Straw Mushrooms, drained and rinsed
100g/4oz cooked crab meat, flaked
100g/4oz mangetout
150ml/1/4 pint chicken stock
1 tbs cornflour
A sprig of coriander, chopped
A dash of soy sauce

In a wok or thick-bottomed pan, heat up 1 tablespoon of the peanut oil until shimmering hot, then throw in the prawns and fry hard for a couple of minutes, until slightly browned. Reserve.

Throw the rest of the peanut oil and the sesame oil into the wok/pan, and fry the garlic, chilli, ginger and spring onion in them for a few minutes. Let the whole thing sizzle nicely.

Add the sherry, sugar, mushrooms, crab meat and mangetout and stir-fry for a few minutes. Take 2 tablespoons of the chicken stock and mix thoroughly with the cornflour. Pour the rest of the chicken stock into the stir-fry along with the coriander and prawns.

While the stir-fry simmers, slowly pour in the cornflour mix, stirring continually. Cook for a further 5 minutes. The mixture should thicken up. Season with soy sauce to taste and serve. This is perfect with plain rice.

'the real key to their appeal lies in their extraordinary texture . . .'

Clear Soup with Enoki Mushrooms

Serves 4

The Enoki is actually a cultivated version of the Winter Mushroom or *Flammulina velutipes*, grown in conditions which make it unrecognizable as the alter-ego of that orangey-brown capped species. It is grown in the dark in glass jars, producing a pale fruit body which elongates itself in vain in the search for light. Its subtle, grape-like flavour is perfect for Oriental soups.

1 litre/1¾ pints light dashi (Japanese fish stock)
2 tbs soy sauce
6 spring onions, chopped
2 tbs mirin or sweet sherry
1 tbs finely chopped ginger
200g/7oz Enoki Mushrooms, separated from one another
300g/10oz tofu, cut into small cubes

To make the dashi, add a 15cm/6in piece of kombu (dried kelp) to the water, bring momentarily to the boil and take out the kombu. Take off the heat and immediately add 15g/½oz dried bonito flakes. Allow them to settle, then gently strain the liquid through a fine sieve or a muslin. Alternatively, use instant dashi, available in Oriental supermarkets.

Heat up all the ingredients in a large pan, apart from the tofu, and simmer for 5 minutes.

Add the tofu, simmer for another 2–3 minutes and serve.

'... grown in the dark in glass jars, producing a pale fruit body which elongates itself in vain in the search for light.'

chapter 5

mushrooms of love

'Mushrooms are a venereous meat,' wrote the Elizabethan philosopher Francis Bacon. Over the centuries, many people have agreed with him, coveting or demonizing a variety of fungi for their supposed aphrodisiac qualities.

In some instances, this has been primarily a matter of shape. Most mushrooms are phallic in general terms, and some are rather overtly so. Given this tendency, it is only natural that their collection and consumption should turn the mind to affairs of the flesh, at least at a subliminal level.

Several of the species featured in this book can be decidedly priapic. We have already alluded to the phallic symbolism of the most prized specimens of the Matsutake, and the same quality is readily perceived among young Fly Agarics and Shaggy Ink Caps. But the king of the penis-like fungi is undoubtedly the Stinkhorn. Its Latin name, *Phallus impudicus*, even means 'the shameless penis'.

The Stinkhorn must have caused quite a bit of blushing on Victorian country rambles. A certain Gwen Raverat has left us an unforgettable image of a stout nineteenth century matron striving to cleanse the Cambridge area of the offending fungus. Interestingly, the matron in question also happened to be Charles Darwin's eldest daughter:

'This was Aunt Etty's greatest invention: armed with a basket and a pointed stick . . . she would sniff her way round the wood . . . then at last, with a deadly pounce, she would fall upon her victim, and then poke his putrid carcass into her basket. At the end of the day's sport, the catch was brought back and burned in the deepest secrecy on the drawing-room fire, with the doors locked, because of the morals of the maids.'

Stinkhorns may have been a bit much for the Aunt Etties of this world, but in other circles they have been looked on much more favourably. In the Middle Ages, for instance, it was widely believed that both God and the Devil had marked certain plants with signs to indicate their potential usefulness to humans, on the principle of 'like affects like'. It didn't take a genius to work out what the Stinkhorn might be good for. This perky mushroom, which spreads its spores by using a strong stench of rotting flesh to lure flies into landing on its sticky head, has been used as an aphrodisiac for centuries.

Although only really edible in its 'egg' stage (and even here we're only talking in relative terms), the Stinkhorn is still occasionally consumed in parts of rural Europe. Until recently, it was even sometimes mixed in with the feed of infertile cattle. Meanwhile in China, herbalists pay high prices for both the eggs and mature specimens of some species of Stinkhorn, once they have been cleansed of slime. Among the most prized is the beautiful, veiled Bamboo Mushroom (*Dictyphora indusiata*) [get picture if possible].

Rather more appealing to the modern palate (and this probably includes anyone you might have your eye on) are those fungi whose reputation rests more on their taste and rarity than on their outline. Chief among these is undoubtedly the Truffle.

Truffles have been linked with sex since ancient times, when they were widely believed to spring from the spilled semen of stags. The extraordinary thing about Truffles is that in their case, the aphrodisiac reputation can actually be scientifically proven. The fact is that their very survival depends on their ability to act as a turn-on for foraging mammals. In order to reproduce, Truffles must first be unearthed. To ensure this happens, they use the oldest trick in the book: sex. Truffles contain a chemical found in the saliva of rutting boars which is irresistible to some mammals, but particularly to female pigs. This aroma accounts for the well-known practice of hunting for Truffles with swine. It also explains why in various parts of the world dogs, bear cubs and goats are used for exactly the same purpose.

Experiments confirm that the pheromone that so excites the sows is also present among humans, in male underarm sweat and in female urine. Men shown photographs of normally dressed women rate them higher on a sexual attractiveness scale when sniffing alpha-andrestenol. Not surprisingly, the chemical is already being added to some expensive cosmetics.

People have been raving about Truffles since time immemorial. They were known to the Babylonians and the Ancient Greeks, and feature in several of the recipes of the great Roman chef Apicius. By the fourteenth century, they were known for their aphrodisiac properties in France, and by the seventeenth we find the first references to their presence in England. Then, in 1825, the celebrated French epicure Brillat-Savarin declared that no meal could be called truly gastronomic if it didn't involve Truffles. A kind of Truffle mania ensued, from which we are still feeling the after-effects.

Truffles are the subterranean fruiting bodies of certain species of fungi, renowned for their intense aromas, and their ability to infuse other foodstuffs with which they come into contact. There are four main varieties: Black, White, Desert and Summer.

LA TRUFFE
Pâtres Piémontais rôtissant des truffes.
Véritable Extrait de viande Liebig.
Reproduction interdite.
Voir l'explication au verso.

black truffles

(*Tuber melanosporum*)

In the Périgord region of France, the area with which it is most closely associated, the Black Truffle (*Tuber melanosporum*) is simply referred to as the 'Black Diamond'. This analogy holds up remarkably well. There definitely is something mineral about this fungus, with its hard black skin covered in pyramidical warts. A good, fresh specimen will even bounce if you drop it on the floor. Then there is the fact that both Truffles and diamonds are mined from the ground. And both of course go for exorbitant prices . . .

The other famous metaphor connected with the Black Truffle concerns chocolates, and here again it is surprisingly apt. Anyone indulgent enough to eat entire specimens raw will be struck by how similar in texture they are to the luxury chocolates that share their name. There is also a direct historical link. Truffles and chocolate both came into vogue in a major way in eighteenth-century Europe. At some point, it was discovered that Truffles could be preserved for a while in port or brandy, to which they would then impart some of their flavour. This liquor could then be poured into a chocolate shell and sealed in, thus combining the two culinary obsessions of the day. The result naturally came to be known as the Chocolate Truffle.

The Black Truffle's celebrated ability to transmit its taste to other foods is shown to excellent effect if it is stored alongside fresh eggs, or if slivers are inserted under the skin of a chicken the night before it is cooked. This propensity is also occasionally exploited by unscrupulous Truffle salesmen, who are able to disguise immature specimens by mixing them in with ripe ones.

Black Truffles cannot be cultivated in the strictest sense of the word – they are always subject to the whim of the weather – but they can certainly be encouraged. Already, around 90 per cent of Périgord Truffles are harvested from the roots of trees that were deliberately inoculated as saplings. In the wild, Black Truffles are particularly associated with oak trees and hazels, but they are also sometimes found growing in conjunction with sweet chestnuts or certain species of pine.

The Black Truffle season in Southern Europe runs from October to March, during which time between 25 and 150 tonnes will be harvested, compared with 2,200 tonnes in 1890.

Black Truffle Omelette

Serves 2

The French, not atypically, prefer the Black ones because they are what grow in France. But it is significant that the Italians, who have access to both varieties, show a marked tendency to rate the White ones more highly. What is beyond debate is the marvellous affinity between both kinds of luxuriously pungent Truffle and the egg. Done correctly, this simple dish is one of the most delicious things on earth.

4 eggs
50ml/2fl oz double cream
Salt and ground white pepper
1 small fresh Black Truffle, brushed clean and finely peeled
1 tbs olive oil

Separate two of the eggs and beat the whites until stiff.

Whisk together the yolks with the two other eggs, the double cream and a little salt and pepper.

Fold in the Truffle slices and the egg whites.

Heat up the oil in a nonstick pan until really hot.

Pour in the mix and beat vigorously for a few seconds. It will bubble up.

Cook on high for a couple of minutes or until cooked, then fold over and serve.

Dover Sole Bonne Femme with Périgord Truffles

Serves 4

This is a traditional French recipe adapted to include Black Truffles. Despite being a genuinely luxurious dish, it is really not that difficult if you have all the ingredients ready-prepared before you cook.

The fish stock

500g/1lb flatfish bones
A sprig of tarragon
1 clove garlic
A sprig of parsley
A glass of white wine
Just enough water to cover the bones
1 shallot, roughly chopped
2 bay leaves

The main dish

2 small Black Truffles
3 shallots, finely chopped
100g/4oz Button Mushrooms, thinly sliced
100g/4oz butter
8 medium-sized fillets of sole (4 come from each fish), seasoned
Salt
White pepper
Flat-leaf parsley to garnish
3 egg yolks

When you buy the sole from the fishmonger, ask for a few extra flatfish bones for the stock. Also get him to fillet and skin the sole.

Simmer all the ingredients for the stock for 30 minutes, then strain off the liquor. There will be some stock left over afterwards, which you can freeze, or chill for use over the next couple of days.

To make the main dish, begin by peeling the Truffles and chopping them into a fine duxelle. Soak them in a little warm fish stock for half an hour.

In a large nonstick pan, gently fry the shallots and mushrooms in half the butter until the mushrooms have released all their liquid and the shallots are soft.

Take the Truffles out of the fish stock, pour stock in the pan with the shallots and mushrooms and boil rapidly. Reduce by about half.

Put the Truffles in the pan and continue to simmer for a few minutes. Reserve this mixture.

Wipe the pan clean, then fry the sole fillets over medium heat in the remaining butter for a couple of minutes on each side. Reserve on a warm plate.

Place the Truffle mixture back in the pan, season with a little salt and pepper and a few leaves of chopped parsley, bring to the boil, turn the heat off, then thoroughly stir in the egg yolks.

Serve immediately with the reserved fillets of sole on plates warmed in the oven.

'... really not that difficult if you have all the ingredients ready-prepared before you cook.'

white truffles

(*Tuber magnatum pico*)

One morning while we were preparing this book, we were sitting in the upstairs office of some friends in the mushroom business when a sudden change came over the room. We all felt a surge of elation, there was a commotion outside, and everyone rose to their feet simultaneously, without quite knowing why. 'Truffles have arrived!' grinned Mike. And all this before they had even been brought into the building . . .

The scent of a fresh White Truffle is so intense that, up close, it no longer really makes sense any more. It seems like something out of a laboratory, and the nose is overwhelmed by the olfactory equivalent of white noise. But from further away, it resolves itself into an aroma that somehow manages to be garlicky, chocolatey and utterly hypnotic all at the same time.

The White or Alba Truffle (*Tuber magnatum pico*) is harvested from late September to early January in the province of Piedmont in north-western Italy. Inevitably, there is a great deal of debate about whether it is superior to the Périgord variety, particularly as the French only produce the latter while the lucky Italians have both. 'The difference between the Truffles of Italy and France is the difference between angels and nuns,' wrote Enza Cavallero, evidently on the side of the angels. The weight of consensus in culinary circles is probably in his favour, particularly when it comes to reaching for the wallet. White Truffles tend to be about three times as expensive as Black ones. A kilo can cost as much as a perfectly serviceable second-hand car.

A less expensive alternative to the Piedmont Truffle is the *Tuber gibbosum*, or Oregon White. This fungus, whose star is very much in the ascendancy, grows in association with the Douglas fir tree on the West Coast of the United States. Americans like to claim patriotically that it can rival anything out of Europe. We'd say the jury is still out on that one.

Because of their intensity, White Truffles are used extremely sparingly. They are often simply grated over egg dishes or pasta, frequently with a special implement which has acquired a cachet all of its own by association.

White Truffle Ravioli with Egg, Spinach and Gruyère

Serves 4

To give you an idea of just how far one skilfully employed White Truffle will go, we refer you to this recipe recently served to Nick at a banquet for two hundred people. One Truffle fed the lot of them. This specimen had been picked up at auction by the organizer for £1,500 and he felt obliged to stage a special charity event to celebrate his purchase. It was the size of a large new potato.

You will need to find a way to slice your Truffle paper-thin. One possibility is to invest in a special Truffle slicer. Another is to use the smallest setting on a mandolin. Alternatively, with practice, you can use a small, sharp knife, or even a razor blade.

The sauce

250ml/½ pint rich chicken stock
3 egg yolks
A pinch of nutmeg
A pinch of ground white pepper
150g/5oz grated Gruyère cheese

The pasta

4 sheets of thin fresh pasta cut into 8 squares or rounds 8cm/3in across

The filling

400g/14oz fresh spinach
25g/1oz butter
A little salt and freshly ground pepper
4 raw egg yolks, superior grade
Paper-thin White Truffle shavings
4 pinches of nutmeg

Make the sauce first. To do this, begin by heating the stock while you mix the other ingredients in a separate bowl. Pour the hot chicken stock over the mix and whisk vigorously.

Place the mixture over low heat for another couple of minutes while continuing to whisk. Reserve, with cling film or greaseproof paper on top.

Cook the pasta until al dente, and reserve.

To make the filling, first blanch the spinach in boiling water. Squeeze out the water, chop finely, then re-heat with the butter, salt and pepper. Have the egg yolks (which must be intact) and the chopped spinach to hand.

Make sure all ingredients are warm, and the plates. Then arrange as follows:

First, place some Gruyère sauce on a medium-sized flat plate. Top this with a layer of the pasta. Next, some chopped spinach, and on top of this, one unbroken egg yolk. Add two or three shavings of the Truffle, then a final layer of pasta. To finish off, top with three or four more Truffle shavings, and a small pinch of nutmeg.

Serve immediately.

Truffle Honey

Serves 4

This is something you will either love or hate. Fortunately, there is no need to break the bank to find out, as you can use peelings from Truffles used in other dishes.

The honey in question should be a mild one, so avoid jungle or pine honeys, or dark varieties in general.

1 pot of clear, mild honey
The peelings from one White Truffle (you could equally well use Black Truffle peelings)

Gently heat the honey, then add the Truffle peelings. Let the honey cool and rest it for a day or two.

Heat the honey up again and strain with a fine sieve. Discard the Truffle peelings, and replace the honey in the pot.

Keep in the fridge and use within 6 months.

White Truffle and Potato Soup

Serves 4

This is a delicate soup, simple to make but very, very expensive. It may seem profligate to combine Truffle with the humble potato, but only until you have tasted this recipe.

On reading the ingredients, Johnny's perfectionist and now deceased grandfather would probably have made someone drive into town to buy him a walnut.

100g/4oz finely chopped leeks
100g/4oz finely chopped shallots
50ml/2fl oz olive oil
1.1 litres/2 pints rich chicken stock
400g/14oz potatoes, peeled and sliced
1 walnut-sized White Truffle, brushed free of dirt
200ml/7fl oz double cream
6 egg yolks
Salt and ground white pepper

Fry the leeks and shallots in the olive oil over medium heat until soft. Pour in the chicken stock, bring to the boil and add the potatoes. Simmer for half an hour, or until the potatoes are slightly overcooked.

Pass the mixture through a conical sieve. Push the soup through with a wooden spoon until you have squeezed every last drop out into a clean pan. (Conical sieves are ideal for this because they are built sturdily to deal with thick liquids. If you don't have one you can use a mouli or a stick blender, but you won't end up with such a velvety texture.)

Use a Truffle slicer to get 12 to 16 paper-thin slices of Truffle. Then finely chop the rest of the fungus.

Heat up the soup to boiling point, then turn the heat down low. Add the chopped Truffles and simmer for a few minutes, then pour in the cream and whisk in the egg yolks and a little salt and pepper. Carry on whisking. After a couple of minutes the soup will thicken slightly.

Serve immediately, with the Truffle shavings on top.

desert truffles

(*Terfezia gp.*)

Various species of Truffle grow in arid or semi-arid parts of the world, including the Sahara, Kalahari and Arabian deserts. During the Gulf War, some Kuwaitis were more concerned about the loss of their beloved Desert Truffles than they were about the oil.

The most important Middle Eastern varieties are the Black and Brown Kame (*Terfezia bouderi* and *claveryi* respectively). These may lack the pungency of their European equivalents, but they are much coveted all the same. The Bedouin have long associated them with thunderstorms, and the Iraqis used to hunt for them by feeling in the sand with their big toes.

summer truffles

(*Tuber aestivum*)

The Summer Truffle (*Tuber aestivum*) does not have the glamour of its sophisticated European relatives, but it is certainly worth eating and seeking out. In any case, in Britain, it's the only option available. Black on the surface and creamy-grey on the inside, the Summer Truffle has a warm, nutty flavour which steadily improves as its season progresses (late summer to mid-autumn).

As with the Périgord variety, the presence of the Summer Truffle is often betrayed by a tell-tale patch of bare ground directly above where it is growing. This 'scorched earth' effect is the result of the fungus depriving the soil around it of nutrients to support the growth of anything else. A friend of ours is convinced he has spotted a Truffle tree in Clapham Common, and we know for a fact that they used to grow in Regent's Park. Not that this is a carte-blanche to head for the Royal Parks with a spade . . .

Duck Liver Pâté with Truffles

Serves 4

Perhaps because we were writing this book, this winter Nick found himself deluged with Truffles. Black ones, White ones, Summer ones from the antipodes and even Chinese ones. To help cope with this unexpected glut, he devised this recipe to prolong the pleasure. The sweetness of the peppers makes a vivid contrast to the pungency of the Truffles.

This pâté will keep in the fridge for up to a week.

1 large yellow pepper
1 large red pepper
400g/14oz duck livers (or use chicken livers)
2 cloves garlic, finely chopped
250g/9oz unsalted butter
A splash of brandy
1 tsp mustard powder
Salt, preferably sea salt
Black pepper
1 tsp finely chopped thyme
50ml/2fl oz double cream
2 Black or Summer Truffles, peeled
100ml/4fl oz beef stock
1 leaf of gelatin

Roast the peppers in the oven for half an hour at 230°C/450°F/gas 8. When you take them out of the oven, put them in a plastic bag to cool. Then peel the skin off, de-seed, and cut into thin strips.

Fry the duck liver and garlic in 50g/2oz butter over medium heat until cooked through. This will take about 8 to 10 minutes.

Take the livers out with a slotted spoon and place in a blender. Wash out the pan with the brandy, and add this to the blender.

Melt the rest of the butter in the pan and pour into the blender. Add the mustard powder, salt, pepper, thyme and double cream.

Blend until smooth. Then pour into a mixing bowl.

Throw in the peppers and mix with a spoon until evenly distributed.

Pour the mixture into a mould lined with cling film, then press down firmly. Cover and leave in the fridge for 24 hours.

Thinly slice the Truffles with a Truffle cutter.

Take out the pâté from the fridge. Turn it upside down and take off the cling film, and completely cover the pâté with the Truffle slices.

Heat up the beef stock, then crumble the gelatin into it. Stir until dissolved.

Paint the stock and gelatin mixture evenly on to the pâté.

The result should be a glistening work of art. Serve it in thin slices with toast.

Truffle hunting with trained animals may be the best known location technique, but it is only one of several. It does have the advantage of ensuring that only ripe specimens are harvested, since immature Truffles fail to elicit a response from the creatures in question. Cruder methods, such as raking the ground in promising spots, are far less discriminating. On the downside, there is always a risk that the animal employed will try to eat the Truffles, particularly if it is a pig. For this reason, many Truffle seekers prefer to use dogs, even though, unlike pigs, they need to be specially trained for the task.

In Sardinia, people search for shallow-lying *Terfezia* varieties by prodding the ground in likely areas with needles mounted on sticks, feeling for a tell-tale resistance. The Bushmen look out for little raised piles of earth. But perhaps the most fascinating of all the many ways of looking for Truffles involves the use of an insect. If an experienced hunter lies down and starts gazing horizontally along the ground, he is probably looking for a female Truffle Fly. These creatures lay their eggs directly above individual Truffles, to ensure their larvae have the shortest possible journey to reach their nursery food. Before they do so, they hover for a time over the chosen spot to fix their bearings. If the hunter digs down from exactly that point, he has an excellent chance of finding a Truffle.

There really is something seductive about Truffles. In part, this is a matter of their inaccessibility: they genuinely are a kind of hidden treasure. It is also, as we have seen, partly a matter of biochemistry – just ask a female pig. Above all, though, anyone prepared to go to the trouble and expense of procuring them for you is definitely serious. They have to be worth a second glance at least.

chapter 6

oddballs

In 1993 the athletics world was stunned by a series of astonishing performances by Chinese female competitors at the Chinese National Games in Beijing and then at the World Championships in Stuttgart. When the inevitable eyebrows were raised, their trainer, Ma Junren, explained that the athletes in question had been on a special diet of *Cordyceps sinensis*, a charming little fungus which grows on the living body of a Tibetan species of caterpillar. Welcome to the margins of the mushroom world. To the fungal Far Side, if you will.

The Kingdom of Fungi is extraordinarily diverse, and has managed to penetrate almost every conceivable ecological niche. There is even, for instance, a species which only lives in rocket fuel. Over the centuries, people have learned to exploit this diversity in all kinds of novel ways. Three examples will give you a flavour:

One little-known property that certain species of fungus possess is bioluminescence, or the ability to glow in the dark. In 1652 the Swedish historian Olaus Magnus described the inhabitants of the far north of Scandinavia strategically placing pieces of a rotten bark on their way into the forests to guide them back home again at night.

Other societies have used fungi to make alcohol. At one level, all fermentation depends on the activity of fungal organisms such as yeasts, but the Aracunas of Chile utilize a species built on an altogether different scale. This is *Cyttaria darwinii*, the Patagonian Gall Fungus. This extraordinary, golf ball-shaped fungus is a natural parasite of the Southern beech tree, with a peculiarly high sugar content which suits it admirably to fermentation. Meanwhile, the strangely scarred wood it leaves behind on its host can be used to make exquisite coffee tables.

In southern Italy, town dwellers used to keep 'Mushroom Stones' in flowerpots. These hard masses were actually filaments of the edible *Polyporus tuberaster*, densely packed together to form what is technically known as a sclerotium. This might include within its mass pieces of earth, pebbles and anything else it had picked up while it was forming. When watered, the 'stone' would produce mushrooms again and again in an apparently miraculous fashion.

Leaving these more exotic uses aside, together with those familiar ones associated with edibility, the most important contributions fungi have made to humanity are undoubtedly medical. In China, and the Orient in particular, where the distinction between food and medicine is far more blurred than in the West, mushrooms have always been taken for their health benefits.

The most revered medicinal mushroom in China is the *Ganoderma lucidum*, or Reishi. Ancient Chinese alchemical texts refer to a fungus called 'Chih', said to bestow long life and almost certainly to be identified with the Reishi. The Emperor Shih Huang Ti (259–210 BC) went to extravagant lengths to search for this wondrous mushroom, despatching an entire fleet to search for it in the islands of the East, but without success. A century later, the Emperor Wu repeated his attempt and repeated his failure. Then, incredibly, in 109 BC, the fungus started sprouting spontaneously in the grounds of his palace. The imperial records for that year describe the springing up of a nine-stalked 'fungus of immortality'.

In modern China, the Reishi or Ling Zhi is used to treat everything from anorexia to heart complaints. It has produced particularly striking results in patients with cancer. After thousands of years of searching for a reliable source, the Reishi was finally successfully cultivated in the early 1970s.

In the light of all that has been learned about the Reishi and the discovery of other remarkable effects among mushrooms used in traditional Chinese medicine, we should perhaps not be too sceptical about the Caterpillar Fungus story. Studies have shown that a regime of *Cordyceps sinensis* can increase the oxygen efficiency of mice by up to 50 per cent. Even the familiar Shiitake has a demonstrably beneficial effect on cholesterol levels. Perhaps, given how deeply mushrooms are integrated into the way of things, we shouldn't be so surprised.

The final group to be considered here, and the crucial one from the point of view of a recipe book, are those fungi that look extraordinary or grow in outlandish fashions, but make for excellent eating all the same. Happily, some of the more outlandish fungi are also among the tastiest. Many oddballs also have the advantage of being almost impossible to mistake, providing an unaccustomed degree of security for the anxious consumer. The remainder of the recipes will revolve around some of these.

giant puffballs

(*Langermannia gigantea*)

'. . . were every spore in a medium-sized Giant Puffball to germinate successfully, after just two generations, the grandchildren would form a pile roughly the size of the sun.'

Giant Puffballs (*Langermannia gigantea*) are the pumpkins of the fungal world – we first encountered them in Chapter Two. If conditions are right, they can grow colossal.

It has been calculated that were every spore in a medium-sized Giant Puffball to successfully germinate, after just two generations, the grandchildren would form a pile roughly the size of the sun. Fortunately, the Giant Puffball is extremely profligate with its spores. It is found in meadows and pastures, often near hedges, and looks from afar like the clean-picked skull of a huge mammal.

One year, our friends in the business were approached before the Puffball season had got started by a man carrying a couple of perfect specimens under his arms. They had, he explained, suddenly materialized in his greenhouse. 'Any good to you?' he asked. They certainly were – Giant Puffballs are much coveted as display items in upmarket restaurants and greengrocers, and, at the time, these were probably the only ones in England. To add to the joy, the man reappeared the following year, and the one after. Somehow, Giant Puffball mycelium must have got mixed up in his compost.

The only real drawback with the Giant Puffball is that some specimens are simply too big for normal-sized groups of people to get through before they start to go off. Johnny's scientifically minded Great-Uncle Guy used to get round the volume problem by slicing just the top few inches off the Giant Puffballs he came across, leaving the bulk of them still attached to the ground. When he returned for another slice, the mushroom would often still be fresh.

Giant Puffball Soufflé

Serves 4

Giant Puffballs look rather as though they have been inflated with air, so it seems only fitting to make them into a dish which really has been. This soufflé also takes full advantage of their very delicate flavour.

75g/3oz butter
Half a Galia melon-sized puffball, peeled and finely chopped
2 tbs plain flour
75ml/3fl oz milk
75ml/3fl oz double cream
4 eggs, whites separated from yolks
30g/1oz grated Parmesan
A pinch of ground nutmeg
Salt and freshly ground black pepper

Heat up the oven to 200°C/400°F/gas 6.

Grease a 1 litre/1 3/4 pint capacity soufflé dish with a little of the butter.

Take the rest of the butter and heat it up in a thick-bottomed saucepan over a medium flame until it starts to bubble.

Add the finely chopped Puffball and fry for 10 minutes or so until cooked. Then blend in a food processor until smooth.

Replace the purée in the pan, add the flour and beat into a paste. Cook this paste gently for about 5 minutes, stirring frequently so it doesn't stick or become lumpy.

Whisk in the milk and cream, adding slowly to prevent lumps from forming. Cook on for 5 to 10 minutes over low heat, until the sauce thickens.

With the heat off, add the egg yolks, Parmesan and nutmeg, and season to taste. Leave the mixture to cool.

Place the egg whites in a large, grease-free mixing bowl (any grease will prevent them from fluffing up properly). Add a pinch of salt to stabilize the egg whites. Then beat with a hand whisk or electric mixer until stiff. When removing the whisk the egg whites form soft peaks which don't collapse under their own weight, they are ready.

Add a couple of spoonfuls of beaten egg white to the sauce and mix in with a rubber spatula.

Take the rest of the egg whites and fold in very gently, keeping as much air in the dish as possible.

Place this mixture in the soufflé dish and cook in the oven for half an hour.

If you are not sure whether it is cooked or not, test with a skewer. If this comes out clean, the soufflé is ready.

Serve with a tossed green-leaf salad.

Puffball Soup

Serves 4–6

Puffballs are perfect for soup: they are soft and therefore blend down nicely, and their delicate flavour makes for a classy broth. An advantage is that you can freeze the end product.

A couple of heaped tbs butter
1 small- to medium-sized puffball, firm and creamy, no sign of yellowing, sliced and diced
2 cloves garlic, chopped
100g/4oz celery, sliced
100g/4oz leek, sliced
100g/4oz onion, sliced
1.25 litres/2 pints chicken stock
300g/10oz potatoes, peeled and sliced
2 tbs chopped dill
Salt and freshly ground pepper
200ml/7fl oz crème fraîche

Melt the butter in a large heavy pan. Add the puffball, garlic, celery, leek, onion and fry for at least 15 to 20 minutes, until the puffball has released its juices and the vegetables have softened.

Add the chicken stock and the potato. Simmer for 40 minutes, until the potato is falling apart.

Add 1 tablespoon of chopped dill, salt and pepper and the crème fraîche. Blend until smooth.

Garnish with chopped dill. This soup would be delicious with toast or croûtons.

chicken of the woods

(*Laetiporus sulphureus*)

There are two important pieces of advice with Chicken of the Woods (*Laetiporus sulphureus*). First of all, only eat the young ones. If one in five of the Chickens you come across in your lifetime are tender enough to eat, you're doing very well. You will be able to tell by the vividness of the colours. Secondly, don't make the mistake of eating specimens that grow on yew trees. They are able to absorb their hosts' toxicity.

Duck, Corn and Chicken of the Woods Stew

Serves 4

Somewhere outside a major European capital there is a forgotten plum orchard – wood rotted, branches splayed, heavy with wasp-infested fruit. We have found magnificent specimens of Chicken of the Woods there, clinging to the tree trunks like weird yellow seeping limpets.

Once the Chicken of the Woods has taken up the flavour of the duck stock, it is difficult to tell it apart from the actual meat in this stew. For this dish it is important to use a whole duck for the flavour. If you don't feel too confident about cutting it up, get your butcher to do this for you. What you want is each leg trimmed and cut into two, plus two breasts (with the wings still attached) also trimmed and cut into two. This should leave you with eight portions of duck, plus the carcass, giblets and trimmings, which you should keep separately for the stock.

1 duck, portioned as above, carcass, giblets and trimmings separate from meat
2 carrots, sliced, plus another roughly chopped carrot for the stock
2 sticks of celery, sliced, plus 1 extra stick for the stock
1 medium onion, chopped, plus another half for the stock
Salt and freshly milled black pepper
300g/10oz Chicken of the Woods, chopped
3 cloves garlic, chopped
4 corn-on-the-cob, chopped into chunky cartwheels
1 red pepper, roughly diced
500g/1lb new potatoes
A couple of sprigs of thyme
2 bay leaves
A small bunch of flat-leaf parsley
4 tomatoes, roughly chopped
A sprig or two of tarragon
A glass of white wine

Simmer the duck carcass, giblets and trimmings together with the stock vegetables for an hour or so in just enough water to cover. Then take the vegetables and duck out, skim the fat off the top with a ladle, and reserve. (There will be a fair amount of fat.) Strain off the stock.

Season the portions of duck with a little salt and pepper and bake in a hot oven (230°C/450°F/gas 8) for 15 minutes.

Fry the Chicken of the Woods, chopped onion, sliced celery, sliced carrots and garlic in the duck fat in a thick-bottomed saucepan until soft. This should take about 10 to 15 minutes. Pour in the duck stock and the rest of the ingredients, after removing the fat which seeped from the portions of duck while they were cooking.

Simmer the stew for an hour or so, making sure it doesn't boil. If it looks as though there isn't enough liquid, top up with water.

At the end of cooking, season with a little salt and pepper and skim any excess fat off the top. Serve with thick wedges of freshly baked bread.

cauliflower fungus

(*Sparassis crispa*)

This bizarre-looking creature (*Sparassis crispa*) is sometimes to be found in the autumn at the base of conifers. It is home to all manner of earwigs and grubs, which must be painstakingly removed if it is to get past most palates.

The Cauliflower Fungus has a unique, jelly-like texture, and a delicate but intensely mushroomy taste. Only young specimens are really worth eating, as the older ones tend to be too dry.

'And thus it was that these unlikely sounding bedfellows made it on to the same plate.'

Lobster Ravioli with Cauliflower Fungus and Saffron Butter

Serves 4

If you are ever lucky enough to find a good Cauliflower Fungus, the occasion merits pushing the boat out a bit. It would be hard to deny that the recipe below fits that description.

Nick created this lobster ravioli for the first class of a major airline, minus the Cauliflower Fungus. However, during the development phase of this air-born dish, he went foraging in Scotland with Dick Peebles, the 'Wild mushroom man of the North'. One of the prize trophies of the trip was a beautiful specimen of Cauliflower Fungus. And thus it was that these unlikely sounding bedfellows made it on to the same plate.

All the constituent parts of this dish can be prepared in advance (either buy live lobsters and poach them in boiling water, or buy the cooked meat frozen) – so don't get stressed out.

The saffron butter

150g/6oz unsalted butter
50ml/2fl oz olive oil
A pinch of saffron strands
A sprig of bush parsley, finely chopped
Ground black pepper and sea salt
Juice of a lemon

The ravioli

1 clove garlic, finely chopped
75g/3oz leeks, finely diced
75g/3oz shallots, finely diced
50ml/2fl oz olive oil
100ml/4fl oz double cream
A sprig or two of bush parsley, chopped
200g/7oz cooked peeled prawns, puréed in the food processor
200g /7oz cooked lobster tail meat, roughly diced
The zest of a lemon
A pinch of saffron strands
Sea salt and ground black pepper
A large thin sheet of freshly rolled pasta
1 egg, lightly whisked

The other bits

4 lobster claws (if you feel confident enough, remove the shell)
4 slices of fresh young Cauliflower Fungus, inspected for vermin
butter for frying

To make the saffron butter, mix together all the ingredients in a food processor, then reserve.

To make the ravioli, gently fry the garlic, leeks and shallots in the olive oil until soft, then add the double cream and boil for a few minutes until reduced by half.

Add the parsley, prawns, lobster, lemon, saffron, salt and pepper, and simmer together for 10 minutes or so. After cooking, this mixture should be damp, but not swimming in liquid, or the ravioli will suffer. Leave this sauce to cool.

Cut out the desired ravioli shapes with a cutter. An 8–10cm/3–4in diameter fluted round cutter would be ideal. Place a mound of cool lobster mix in the middle of half of the ravioli shapes, leaving a generous margin clear at the sides.

Paint the periphery of the pasta with egg wash, then gently lay another sheet of ravioli on top. Firmly press the sheets together, starting right up next to the mound of fish mixture. (Try to eliminate all the air, or the ravioli will explode when you cook them.)

Steam the ravioli for 4–5 minutes, or simmer in a little water for the same length of time.

Heat the saffron butter up in a saucepan with the lobster claws.

Season the Cauliflower Fungus slices with salt and pepper and fry in butter for 3 to 4 minutes on each side until nicely browned.

Serve with each portion consisting of one ravioli on a bed of Cauliflower Fungus, with a lobster claw perched on top. This mini-construction should then be drenched with the saffron butter.

jew's ear fungus

(*Auricularia auricula-judae*)

According to legend, after he had successfully betrayed Jesus, Judas Iscariot hanged himself on an elder. As a result, the peculiar ear-shaped fungus which grows on that tree became known as 'Judas' Ear. In time, this was corrupted to 'Jew's Ear'. Ironically, in traditional medicine it is usually prescribed for throat complaints.

The jelly-like texture of this mushroom goes down particularly well in China, where feel in the mouth is rated as highly as taste. When dried and subsequently rehydrated, Jew's Ears can swell to a remarkable size.

Given a spell of warm, humid weather, the Jew's Ear can appear at any time of the year. As usual, the young ones are the best.

'As usual, the young ones are the best.'

Chinese Restorative Jew's Ear Fungus

Serves 4–6

The Chinese make little distinction between food and medicine. If you go to a Chinese herbalist, he is likely to give you a small bundle of ingredients and tell you to go off and make soup! Next time you have a cold or the 'flu, you might want to try doing exactly that. This warming broth is designed (rather paradoxically) to reduce the heat in the body, thereby restoring yin yang balance. It is also full of interesting textures and flavours, and very simple to make, so long as you aren't in a hurry.

Don't be daunted by the 'Chinese chicken soup herbs' listed among the ingredients. Any Oriental grocer is likely to know exactly what you mean.

A couple of handfuls of dried Jew's Ear Fungus
1 large boiling chicken
2 litres/3½ pints water
125g/5oz Chinese chicken soup herbs (available from any Chinese grocers)
2 x 2.5cm/1in pieces of root ginger
500g/1lb pak choi, washed and roughly chopped
Soy sauce, to taste

Soak the Jew's Ear Fungus in warm water for half an hour, then rinse and finely slice. Reserve.

Place the chicken in a large stockpot with the water, add the herbs and ginger and simmer for around 4 to 6 hours, skimming occasionally. Do not boil.

Strain the stock and return to the pan, reserving the chicken meat. Add the Jew's Ear Fungus to the stock and simmer for an hour.

Add the chicken meat, which should be shredded.

Add the pak choi and the soy sauce, simmer for just a minute or so and the soup is ready to serve.

amethyst deceiver

(*Laccaria amethystea*)

It is not quite clear who or what the Amethyst Deceiver (*Laccaria amethystea*) is supposed to deceive but it undoubtedly has a beautiful colour. The Deceiver is more of a novelty item than anything else. As it is virtually scentless and tasteless, you can pretty much use it as you wish. Hence its potential presence in desserts.

The Amethyst Deceiver grows in the woods, often near beech trees, from late summer to early winter. When you have eaten every other kind of wild mushroom, why not try one of these?

Tomato Soup with Amethyst Deceivers

Serves 4–6

This is a wonderful showcase for a colourful mushroom, as it floats on a deep red sea of tangy tomato soup.

600g/1¼ lb ripe red tomatoes
100ml/4fl oz olive oil
Salt and black pepper
100g/4oz carrot, roughly chopped
100g/4oz celery, roughly chopped
100g/4oz onion, roughly chopped
100g/4oz bacon, or bacon rind, roughly chopped
2 cloves garlic, roughly chopped
1 litre/1¾ pints chicken stock
1 large potato, peeled and roughly chopped
A sprig of thyme
A sprig of parsley
2 bay leaves
100 ml/4fl oz double cream
A couple of handfuls of Amethyst Deceivers

Roast the tomatoes in the oven for 30 minutes at 220°C/425°F/gas 7 with a little olive oil, salt and pepper.

Gently fry the carrot, celery, onion, bacon and half the garlic in some more olive oil, until the vegetables soften.

Add the chicken stock, potato, thyme, parsley, bay leaves and simmer for half an hour. Then add the tomatoes.

Pass through a mouli, or blend with a blender. (If you do the latter, the soup will be less velvety, and you should take out the bay leaves, thyme and bacon rind before you blend.)

Place back on the heat, add the cream and season.

Finally, fry up the Amethyst Deceivers with the rest of the garlic in a little olive oil, and scatter on top of individual portions of soup.

essentials

While it would have been negligent of us not to include some information on mushroom identification and preparation regarding the species featured in the recipes in Chapters One to Six, we must emphasize that the following 'Cook's Checklist' should be taken only as a starting point. Readers should seek corroboration not only from a specialized identification book, but – more importantly – from a suitably qualified individual. In Continental countries the local pharmacist often performs this function officially. In Britain finding a competent mycologist is a more hit-and-miss affair. Contact your local naturalist groups, or one of the organizations listed on page 158.

REMEMBER, NEVER EAT ANY MUSHROOM THAT YOU CANNOT POSITIVELY IDENTIFY.

cook's checklist

The mushrooms in the following Checklist are listed alphabetically by official botanical name. As with garden plants, mushrooms keep being reclassified by experts. We have listed the mushrooms according to the Latin names with which we are most familiar. (You may find them under different names in older books or books published overseas, or used by maverick authors and experts who disagree with official classifications.) Hence the 'aka' or 'also known as' line, which gives synonyms where they exist – technical niceties that equip mushroom enthusiasts with a lingua franca for pursuing their quarry in all countries and all languages. You will find both the synonyms and the various common names cross-referenced in the Index.

Naming is one thing, but we repeat:

NEVER EAT ANY MUSHROOM THAT HAS NOT BEEN POSITIVELY IDENTIFIED.

The seasons listed for the various species are based on the typical timings of their appearance in temperate parts of Europe. Generally speaking, the greater the distance from the equator the shorter the season and vice versa. Much of the USA experiences similar growing seasons to those in Europe, but in the Gulf States of America fruiting may occur right through the winter, and on the West Coast it is often determined more by precipitation than conventional 'season'. In the southern hemisphere, species will usually grow in the same seasons as their counterparts in the north but at opposite times of the year.

agaricus

Also known as *Psalliota*

To most unenlightened Anglo-Saxons, these are THE edible mushrooms – all the rest are dismissed as toadstools. Both wild and cultivated *Agaricus* may be recognized by their classic mushroom shapes and gill arrangements. Several *Agaricus* species and varieties are edible and good, but beware of any that stain or bruise bright yellow.

agaricus bisporus

Common names: Cultivated or White Button Mushroom, Champignon de Paris.

Description: White to medium-brown caps. Pinkish gills darkening brown with age. The cultivated cousin of the Field Mushroom.

Habitat: Supermarket shelves.

Season: Year round.

Identification Tips: Read the label!

Preparation: Wipe or brush clean in preference to washing with water. Stalks and caps are both edible, and this is one of the few mushrooms that is good raw.

Storing and Preserving: Shops often provide brown paper bags for loose cultivated mushrooms, which are ideal for transporting them and for keeping them in the short term. If they come in a container with a cling-film lid, punch a few holes in it to let them breathe. *A. bisporus* can be frozen after blanching or frying.

Notes: The Brown or Crimini Mushroom is simply a darker strain of *A. bisporus*. Open-capped examples are known as Portobello Mushrooms.

Left: *Agaricus bisporus*

agaricus campestris

Common Name: Field Mushroom.

Description: White caps. Deep pink gills, darkening brown with age. Traditional mushroom smell, just like supermarket mushrooms. Flesh bruises very slightly pink.

Habitat: Fields and pastures.

Season: Late summer to autumn.

Identification Tips: Small, simple ring on the stem.

Preparation: Both stalks and caps are edible, and may be eaten raw. Wipe or brush clean, and look out for maggots.

Storing and Preserving: As for *A. bisporus*.

Notes: Other excellent edible wild *Agaricus* species include the Horse Mushroom (*A. arvensis*), which smells of aniseed, and The Prince (*A. augustus*) which smells of bitter almonds.

amanita caesarea

Common Name: Caesar's Mushroom.

Description: Smooth orangey-red caps, fading somewhat with age.

Habitat: Not yet found as far north as the UK. Grows in open deciduous woodland, particularly near oak trees, in Mediterranean Europe.

Season: Summer to autumn.

Identification Tips: Looks as though it grows out of an egg. This is in fact the protective volval sac by which all *Amanitas* can be distinguished.

Storing and Preserving: Wrap each cap individually in cling film and freeze straight away. Then cook from frozen, without defrosting.

Notes: Some *Amanita* species are highly poisonous (notably the Death Cap (*A. phalloides)* and the Destroying Angel (*A. virosa*)), or hallucinogenic. The Fly Agaric (*A. muscaria*) is the best known example of the latter. In European markets Caesar's Mushroom is commonly sold in its egg stage. A Romanian consignment to the UK was recently found to contain specimens of the Death Cap. Fortunately the mushrooms were identified before anyone came to harm.

Left: *Agaricus campestris*

armillaria mellea

Common Names: Honey Fungus, Honey Mushroom.

Description: Grows in large clumps. Strong smell and bitter taste when raw.

Habitat: Tree stumps and roots.

Season: Summer to early winter.

Identification Tips: It's the one eating your garden. Usually grows in dense clumps. Whitish gills.

Preparation: Eat only the caps. Blanch before use, discarding the water.

Storing and Preserving: Pickles very well.

Notes: Discard the stalks, which are too tough to eat. Some people find Honey Fungus indigestible even when blanched first.

auricularia auricula-judae

Also known as *Hirneola auricula-judae*

Common Name: Jew's Ear Fungus.

Description: This is one of the jelly fungi (*Heterobasidiomycetes*, since you ask). Ear-shaped. Reddish brown colour. Gelatinous when fresh, hardens on drying.

Habitat: Grows on dead branches, particularly of elder and elm.

Season: Year round.

Identification Tips: Virtually unmistakable.

Preparation: The whole thing is edible, but wash thoroughly before use. This fungus needs to be thoroughly cooked before it is eaten, and can take long cooking.

Storing and Preserving: Can be dried and reconstituted. Related species such as the Black Jelly Fungus (*A. polytricha*) are cultivated by the Chinese.

Notes: Look under rotting logs, as they can sometimes be hidden.

boletus

The Boletes, characterized by spongy pores beneath the caps, include some of the classic gourmet mushrooms. Besides the Boletus species mentioned here, other good edible members of the family include *Boletus aereus* and the Orange Birch Bolete (*Leccinum versipelle*).

Boletus edulis

boletus edulis

Common Names: Cep, Penny Bun, Porcini (Italian), King Bolete (American).

Description: Smooth brown caps, which go slightly sticky in wet weather, and which can be up to 25cm/10in across. Robust, whitish stems, often wider than the caps at their maximum.

Habitat: Woods, in association with a wide variety of trees, including oaks and beeches.

Season: Mid-summer to mid-autumn.

Identification Tips: White 'netting' on the stem. Flesh remains whitish when cut.

Preparation: Clean by brushing the caps and trimming the base thoroughly. Always cut down the middle to inspect for maggots. Damaged areas can be cut away provided the rest is firm. Both caps and stalks are edible. Ceps are OK to eat raw.

Storing and Preserving: Dries exceptionally well. Both raw and cooked specimens can be frozen. Ceps can also be pickled or used for duxelles.

Notes: Watch out for maggots. Ceps can have a very brief season, but there is sometimes a second flush.

boletus badius

Also known as *Xerocomus badius*

Common Names: Bay Bolete, Bay Boletus.

Description: Bay-brown cap, darker stem than the Cep.

Habitat: Woodland, often with conifers.

Season: Autumn, typically just after the Cep.

Identification Tips: Yellowish pores which stain blue rapidly when bruised (this doesn't spoil their edibility). Longitudinal streaks on the stem.

Preparation: Clean by brushing/wiping the caps and trimming the bases. All parts can be eaten. We would advise you to cook them rather than trying them raw.

Storing and Preserving: Very adaptable, like Ceps.

Notes: The Bay Bolete is less prone to maggot infestation than the Cep.

calocybe gambosa

Also known as *Tricholoma gambosum*

Common Name: St George's Mushroom.

Description: White. Often grows in fairy rings, which can be enormous.

Habitat: Typically grows on grass.

Season: Spring (ideally 23 April). Seems to be getting both earlier and later. . .

Identification Tips: Mealy, floury smell. The edge of the cap rolls inwards, particularly when young.

Preparation: Brush well. Stalks are sometimes tough and fibrous, but are edible in younger specimens. Not good raw.

Storing and Preserving: Can be successfully frozen if sautéed first.

Notes: The early growing season makes the St George's Mushroom relatively difficult to misidentify.

Left and opposite page: *Boletus badius*

Cantharellus cibarius

cantharellus

cantharellus cibarius

Common Names: Chanterelle, Girolle (French).

Description: Bright orange-yellow. Gregarious.

Habitat: Woodland species, often growing under beech or oak or sometimes with conifers.

Season: Early summer to mid-autumn.

Identification Tips: Apricoty smell. Tend to be vermin-free, although the edges have often been nibbled by slugs or snails. The thickish 'gills' beneath the cap are actually ridges of flesh which tend to run down the stem and branch irregularly.

Preparation: The whole thing is eminently edible, although chefs tend to most prize the Chanterelle in its young, 'button' form. Brush or wipe rather than wash. We wouldn't recommend eating them raw.

Storing and Preserving: Dry them on a thread. Button Girolles are commonly frozen. Big ones are better frozen in a sauce.

Notes: You may need to trim older ones to get rid of grit or damaged areas.

cantharellus infundibuliformis

Also known as *Cantharellus tubiformis*

Common Names: Winter Chanterelle, Autumn Chanterelle, Yellowlegs.

Description: Tend to grow in huge quantities where found.

Habitat: Grows in coniferous or deciduous woodland, with a preference for acid soils.

Season: Autumn, sometimes into early winter.

Identification Tips: The stalks are hollow and usually lighter in colour than the caps.

Preparation: Pull the stalks apart to check for vermin. Do not try to wash them, or they will turn mushy. If you plan to use them relatively soon after picking, store them in the fridge on kitchen paper. This will absorb the moisture they release. Both stalks and caps are edible, but not good raw.

Storing and Preservation: Best made into sauces and frozen. Can be pickled or stored in alcohol.

Notes: Winter Chanterelles are most easily picked with scissors.

coprinus comatus

Common Names: Shaggy Ink Cap, Lawyer's Wig.

Description: Distinctive curling scales on the cap, hence the name.

Habitat: Likes recently disturbed soil. Often grows on roadsides, new lawns and rubbish tips.

Season: Late summer to autumn.

Identification Tips: Rapid deliquescence.

Preparation: Pick young specimens whose gills are still white and whose caps are yet to open. Do not wash, as Shaggy Ink Caps are extremely fragile, but clean by brushing gently. The whole fruiting body is edible, but not worth trying raw.

Storing and Preserving: Eat straight away (although flavour can be preserved in broth).

Notes: The Common Ink Cap (*Coprinus atramentarius*) is edible and good, but it **MUST NEVER BE COMBINED WITH ALCOHOL.**

Right: *Coprinus comatus*

craterellus comucopioides

Common Names: Horn of Plenty, Black Trumpet, Death Trumpet, Trompette de Mort (French).

Description: Hollow, funnel-shaped fruit bodies, ranging in colour from greyish to dark brown/black, depending on weather conditions.

Spore Deposit: White.

Habitat: Grows in deciduous woodland.

Season: Late summer to late autumn.

Identification Tips: Hard to find, well-camouflaged by its colour.

Preparation: Brush rather than wash clean, as they are very delicate. Check carefully for insects (see below). The whole fungus is edible, but trim the bases before use. Not good raw.

Storing and Preserving: Can be dried, pickled and frozen if cooked first.

Notes: Because of their shape, bugs often crawl inside, so it can be a good idea to tear them into strips before use to expose any creepy-crawlies.

Craterellus comucopioides

flammulina velutipes

Common Names: Winter Mushroom, Velvet Shank.

Description: Smooth, slimy, tan-coloured caps, growing in clusters.

Habitat: Often grows on elm.

Season: Late autumn to spring. Unfazed by frost.

Preparation: Eat only the caps of the wild form; cook thoroughly to overcome their toughness.

Notes: In its domesticated form, where it is grown in the dark to produce a pale, very different-seeming fruiting body, *F. velutipes* is known as the Enoki Mushroom (see page 94).

ganoderma lucidum

Common Names: Varnished Polypore, Reishi (Japanese), Ling Zhi (Chinese).

Description: A stemmed bracket fungus, with characteristic glossy, 'varnished' top surface. Concentric bands of colour, lightening towards the margin.

Habitat: Grows on the roots of deciduous trees.

Season: Year round.

Storing and Preserving: The Chinese dry this fungus before grinding it into a powder.

Notes: This fungus is not edible in the conventional sense, its primary application being medicinal. It is usually used in powdered form. It is a highly important medicinal species in the Orient.

Hydnum repandum

hydnum repandum

Common Names: Hedgehog Fungus, Wood Urchin, Pied de Mouton (French).

Description: Velvety, cream-coloured caps with distinctive spines rather than gills underneath.

Habitat: They like dark, damp places such as ditches. Usually with oak.

Season: Late summer to late autumn.

Identification Tips: Come up in same spot year after year.

Preparation: You may want to remove the spines before use for aesthetic reasons (they are likely to break off during cooking). The whole mushroom is edible and, if you insist, can be eaten raw.

Storing and Preserving: Can be dried or pickled, or blanched and frozen.

Notes: Eat only young ones, as older specimens tend to taste bitter.

laccaria amethystina

Common Name: Amethyst Deceiver.

Description: Vivid lilac all over.

Habitat: Woodland, both deciduous and coniferous.

Season: Late summer to early winter.

Identification Tips: The colour is the key.

Storing and Preserving: Can be stored in vodka, apparently.

Notes: The fruiting bodies of this species are very variable in appearance. The colour is more consistent. Beware of confusion with lilac-coloured species of *Cortinarius*.

lactarius deliciosus

Common Name: Saffron Milk Cap.

Description: Crunchy, brittle texture. Bright orange 'milk', which gradually turns green on exposure to air. Bruises easily, leaving verdigris marks. These are unsightly, but don't affect the taste adversely.

Habitat: Grows under conifers, especially pines and spruces.

Season: Summer/autumn.

Identification Tips: Green bruising.

Preparation: The widely spaced gills provide plenty of opportunity for dirt to get in, so clean thoroughly. The whole mushroom is edible. Blanch first if pickling. They pickle particularly well.

Notes: It can be worth sweeping away pine needles in suitable spots to see if there are any Saffron Milk Caps submerged. If you find any, handle them gently to minimize the risk of bruising.

Left: *Laccaria amethystina*

laetiporus sulphureus

Common Names: Chicken of the Woods, Sulphur Polypore.

Description: Bracket fungus, vivid yellow/yellow-orange when young, whitening with age.

Habitat: Grows on oaks, fruit trees, ash and yew.

Season: Spring to autumn.

Identification Tips: Not easily confused with other species.

Preparation: Must be cooked before eating. Some people find Chicken of the Woods upsets their stomachs. Blanching before cooking minimizes the risk. As the flesh is so dense, it can be washed with impunity.

Storing and Preserving: Not recommended.

Notes: Only young ones are edible. You can tell by their almost luminous colour, which fades as they age. Young ones also release a yellowish juice when squeezed. NB Do not collect specimens from yew trees, however tempting. The host tree is capable of passing on its poisons to its parasites…

Right: *Laetiporus sulphureus*

Left: *Langermannia giganta*

langermannia giganta

Also known as *Calvatia gigantea*

Common Name: Giant Puffball.

Description: Huge, white and loosely spherical. Reminiscent of overgrown marshmallows.

Habitat: Fields, lawns and roadside verges, often among nettles.

Season: Summer to mid-autumn.

Identification Tips: Pumpkin-like dimensions. No stalk.

Preparation: Wipe clean. Cut off any soft or discoloured parts ruthlessly. Do not eat raw.

Storing and Preserving: Not really its forte, unless incorporated into a mushroom stock.

Notes: Eat only specimens that are pure white all through, with no trace of discolouration.

Some smaller Puffball species are also good to eat, among them the very common *Lycoperdon perlatum*, with its distinctive 'goosepimpled' surface.

lentinula edodes

Common Name: Shiitake.

Description: Delicious aroma.

Habitat: Grows on logs. Not found wild in Britain, but now increasingly commercially cultivated.

Season: Year round (at least, the domesticated ones).

Preparation: The stalks can be tough, but they are still perfectly edible.

Storing and Preserving: Excellent dried. Can be cooked and frozen.

Right: *Lentinula edodes*

lepista nuda

Common Names: Wood Blewit, Pied Bleu (French).

Description: Blueish/lilac, particularly on the underside. Browns with age. Can grow in extremely large numbers.

Habitat: Grows in both woodland and gardens. Also increasingly found on supermarket shelves (as Pied Bleu).

Season: Autumn to early winter.

Identification Tips: Strong, perfumed aroma. Poking around in leaf litter may reveal hidden Blewits.

Preparation: Blanch before use, and never eat raw. Both caps and stalks are edible.

Storing and Preserving: Blewits dry and pickle well. Cooked ones can be frozen.

Notes: Some people are mildly allergic to Blewits, but they are good enough to make it worth your while to find out if you are one of them.

The closely related Field Blewit or Blue-leg (*Lepista saeva*), which grows in grassland, is also eminently edible. Indeed, some would say it is tastier.

macrolepiota procera

Also known as *Lepiota procera*

Common Name: Parasol Mushroom.

Description: Gangly stalk, large cap with radiating brownish scales.

Habitat: Open woods and pastures and roadside verges.

Season: Summer/autumn.

Identification Tips: Can be spotted from a long way off – scour the horizon. The ring on the stalk is movable.

Preparation: Discard the stalks. We wouldn't recommend eating the caps uncooked.

Storing and Preserving: Can be cooked and frozen. The stalk is inedible.

Notes: Can grow to more than 30cm/1ft in height and/or cap diameter.

Right: Macrolepiota procera

morchella esculenta

Common Name: Morel.

Description: Distinctive pitted head. Hollow fruit body (both head and stem).

Habitat: Notoriously unpredictable, but Morels are known to like freshly composted, burned and disturbed ground. Sandy soil specialists.

Season: Spring.

Identification Tips: Take care not to confuse with the False Morel (*Gyromitra esculenta*), which is reddish brown and more convoluted in appearance, with a stem divided into chambers rather than being hollow throughout.

Preparation: Inspect internal and external cavities carefully for bugs. Morels can be washed if necessary. The whole fruit body is edible, but **MUST BE COOKED**.

Storing and Preserving: Don't freeze them raw. Otherwise almost anything goes.

Notes: If you find very dirty Morels, they may be better for drying than for using raw.

Above: *Morchella esculenta*

marasmius oreades

Common Name: Fairy Ring Champignon.

Description: Small tan to buff mushroom with a large umbo (bump) in the centre of the top of the cap. Relatively widely spaced gills. The stems are thin, rigid and whitish.

Habitat: Grows in short grass, typically on lawns and pastures.

Season: Late spring to late autumn.

Identification Tips: Grows in rings.

Storing and Preserving: Fairy Ring Champignons dry very well.

Notes: Take care not to confuse with the highly poisonous *Clitocybe rivulosa*, which also grows in fairy rings, but which lacks the umbo (see above) and has crowded gills.

pleurotus

Like *Agaricus*, the genus *Pleurotus* straddles the customary divide between the wild and the cultivated. The culinary value of Oyster Mushrooms perhaps lies as much in their texture as in their taste.

pleurotus ostreatus

Common Name: Oyster Mushroom.

Description: Grows in tiers, with either no stems or very short ones, laterally attached like a bracket. The caps vary in colour from blue/grey to brown. Cultivated varieties come in various pastel shades.

Habitat: Tree stumps and trunks, particularly beech. Domesticated versions are sold in supermarkets.

Season: Year round.

Identification Tips: Pleasant mushroomy odour.

Preparation: Wild Oyster Mushrooms can be maggoty, so inspect carefully. Don't wash them, as they soak up water, but wipe or brush as necessary. We don't recommend eating them raw.

Storing and Preserving: No need to bother, unless freezing a sauce.

Notes: P. cornucopiae is a good close relative which typically lives on elm and oak.

If you find a flush of Oyster Mushrooms in the forest, scrape away the surface of the wood they have grown on to reveal fresh wood for the mycelium to gather nutrients from. This will ensure future flushes.

Left:*Pleurotus ostreatus*

sparassis crispa

Common Name: Cauliflower Fungus.

Description: Unmistakable crimpled appearance. Doesn't look much like a cauliflower, though.

Habitat: Grows at the base of conifers.

Season: Autumn.

Identification Tips: Unmistakable.

Preparation: The convoluted surface is a grit magnet and a paradise for insects. To clean a Cauliflower Fungus properly, you may need to resort to tweezers. Don't try to wash them in water, or they may fall apart.

Storing and Preserving: Try freezing in cooked slices, or in duxelles.

Notes: Don't pick the whole specimen, just cut off a portion. This will allow future growth.

tricholoma matsutake

Common Names: Matsutake, Pine Mushroom.

Description: Amazing scent – spicy, fruity and piny.

Habitat: Pine forests. Virtually identical species are found in Korea and Japan, Northern Sweden and the West Coast of the USA.

Season: Autumn.

Identification Tips: Thick, cotton-like partial veil which breaks as the caps open. Distinctive aroma.

Storing and Preserving: Matsutake can be frozen whole (the Japanese tend to wrap them in tin foil first) or pickled.

Notes: The similar but marginally less prized species *Tricholoma magnivelare* is harvested on the West Coast of the United States.

tuber aestivum

Common Name: Summer Truffle.

Description: Similar in external appearance to the Périgord Truffle, though not growing as large.

Habitat: Grows underground, usually near a beech tree, in chalky soils.

Season: Late summer/autumn.

Identification Tips: When cut, the flesh is whitish, later turning marbled grey.

Preparation; Storing and Preserving: As for Black Truffle.

Notes: Look for tell-tale bare patches around the base of suitable trees. A British native!

Tuber aestivum

Tuber melanosporum

tuber melanosporum

Common Names: Black Truffle, Périgord Truffle.

Description: The surface is covered with pyramidical warts, and the flesh is brownish-black throughout when cut.

Habitat: Grows in association with the roots of oaks, hazels and sweet chestnuts.

Season: Mid-autumn to early spring.

Identification Tips: Use your nose (or one belonging to a suitable animal).

Preparation: Brush well, but never wash in water. If you have to peel them, do so very thinly, and find some use for the peelings! Truffles can be eaten raw, and often are.

Storing and Preserving: Can be found in myriad pastes and oils, embedded in pâté, tinned, bottled, you name it. Just don't bother trying to dry it.

Notes: Black Truffles can be as much as 15cm/6in in diameter.

tuber magnatum pico

Common Names: White Truffle, Alba Truffle, Piedmont Truffle.

Description: Not unlike a smelly, yellowish pebble.

Habitat: Grows underground in deciduous woodland in Northern Italy.

Season: Early autumn to mid-winter.

Identification Tips: Pungent, garlicky aroma.

Preparation; Storing and Preserving: As for Black Truffle.

Notes: The Oregon White Truffle (*Tuber gibbosum*) is increasingly popular in the USA.

Tuber magnatum pico

volvariella volvacea

Common Name: Paddy Straw Mushroom.

Description: Chewy, somewhat slimy-textured grey/black mushroom grown commercially on rice straw.

Habitat: Usually found in tins.

Season: Year round.

Identification Tips: Read the label.

Preparation: Straightforward, assuming (as we shall) that you get them out of a tin.

Volvariella volvacea

poisonous mushrooms

Although it is a central thesis of this book that the likes of the British and the Irish have historically tended to overreact to the threat posed by poisonous mushrooms, the dangers that lurk for the unwary or careless forager are very real indeed.

The French have the comfort of a law which obliges pharmacists to identify any mushrooms which are brought in for their inspection. The rest of us have to take responsibility for identification ourselves.

If you want to be really sure what you are doing, you'll need to invest in a microscope and start identifying mushrooms by their spore characteristics. To take a spore print, place a mushroom cap on a clean piece of glass or sheet of white paper, and cover with a glass to prevent drying out. Leave overnight. You will find the spores under the cap, or, in the case of a spore-shooting species, in a 'halo' around it. However, to learn how to interpret what you find, you will need to refer to a more detailed field guide than this one.

What you should do immediately is learn to identify the three species below, particularly the Death Cap, which is responsible for some 90 per cent of all mushroom-related deaths, and not that uncommon.

Amanita phalloides
Death Cap
Yellow or greenish cap, though sometimes paler. Whitish stalk and gills, large volval sac at the base. Sickly sweet smell.

The toxins in the Death Cap attack the liver and kidneys. The first symptoms are stomach pains, vomiting and an intense thirst some 6 to 24 hours after consumption, followed by a cruelly deceptive period of apparent recovery. A few days later, the victim may die of kidney and liver failure.

Amanita virosa
Destroying Angel
This beautiful, pure white mushroom has a sickly sweet smell, which should be grounds for rejection on its own. It also has a bag-like volva and a shaggy, fibrous stem.

Gyromitra esculenta
False Morel
This is deadly when raw. Although *Gyromitra* are sometimes eaten in Eastern Europe after being thoroughly boiled, it just isn't worth risking.

The convoluted caps are much more irregular than those of true Morels, and the stalks are chambered rather than completely hollow. The best method of identification, however, is spore analysis. *Gyromitra* spores contain at least two droplets of yellowish oil.

collecting wild mushrooms

Debate rages about whether picking wild mushrooms adversely affects their numbers. Some research argues that it positively promotes fruiting, while other experts are more pessimistic. The most important thing is to try to avoid disturbing the mycelium by trampling or other damage. On a different level it is important to try to avoid bringing mushroom collecting into disrepute: be discreet! Don't damage the habitat: in any way, or leave a site littered with mushroom debris.

Overseas countries have a whole range of systems regulating collection, from licensing individuals to restricting communities as to the days on which they may pick fungi and the amounts they may collect.

English Nature publishes a *Code for Collectors*, which offers some useful advice.

Beginners will learn a great deal from going on organized forays – not only about basic identification techniques but about mushroom ecology, likely habitats and so on. Contact local (or national) mycological societies for more information (see addresses on page 158).

Where

All mushroom hunters have their own secret hunting grounds, sometimes, perhaps, illicit ones. We wouldn't be doing our job if we didn't at least mention the legal and/or ethical position. In Britain collecting of any plant (or animal) material from nature reserves, for instance, is vigorously discouraged. Collecting in some areas is severely restricted by local bye-laws. Basically, you are always encouraged to get the landowner's permission. In an ideal world enlightened conservationist landowners MIGHT just realize that fungi are a useful asset, or a valuable crop (there are leaflets available on 'Managing your land with fungi in mind').

Organized forays in Britain will have obtained permission to collect fungi, but sometimes only for identification purposes. Check with the leader whether collecting for the pot is acceptable on a given occasion.

When

Although species vary in their growing seasons, generally speaking the best time of year for mushroom hunting is the early autumn, before the falling leaves have had a chance to cover the emerging fruiting bodies. Ideally, the weather will be warm and humid, and there will have been a decent soaking of rain in the recent past. It pays to keep a careful eye on rainfall patterns, as a good shower will often induce a 'flush' of wild mushrooms. It is also worth bearing in mind that in foraging as in so much else, the early bird tends to get the worm. A good tip is to use your nose when out in the field. Many species are far easier to smell than to see.

Collecting equipment

The following is a basic checklist of mushroom-hunting equipment. Serious enthusiasts may want to keep some of the items about them at all times, just in case.

- A flat-bottomed basket, to minimize your mushrooms rolling around.

- A knife, ideally with a brush attached.

- A notebook and pen(cil) for recording where you find each species (look at surrounding trees and vegetation), and how they look and smell etc.

- Tins, plastic food boxes, or sheets of waxed paper, to use for keeping different species apart in your basket. Plastic bags are a bad idea, as they tend to make mushrooms sweat and accelerate decay.

How

It helps to trim off any dirty parts as you pick the mushrooms, before you put them in your collecting box or basket. Trim away the earthy bases of stalks, for example. Brush any leaves or conifer needles off the tops of caps. Carrying mushrooms with gills downwards, or wrapping oddly shaped ones like Morels and Cauliflower Fungus, prevents dirt from falling into awkward crevices.

Distinguish between fruit bodies with superficial damage, such as nibbling by animals competing for your meal, and ones that are past their prime and beginning to deteriorate. You can trim away external damage, and even internal parts containing maggoty 'passengers' (according to how hungry and/or squeamish you feel). However, once the flesh has begun to decay, even the most reliable edible mushrooms can contain dangerous toxins.

Keep picked mushrooms as cool as possible. Leaving them in a hot car (like carrying them in a sweaty plastic carrier bag) is a recipe for deterioration. Sort them through again as soon as you get home.

Avoid picking beside busy roads (the habitat 'roadside verges' means quiet country roads). Apart from the obvious dust and dirt generated by traffic, mushrooms are classic accumulators of heavy metals from the atmosphere...

And on this cheerful note...

preserving mushrooms

There are numerous ways of preserving wild mushrooms. This section covers the main six. Other techniques include powdering dried mushrooms to use as a condiment, and conserving them in oil.

Drying

Not all species of mushroom dry well: with Puffballs and Ink Caps, for instance, you can forget it. But some varieties positively improve upon drying. Chanterelles, Morels, Ceps and Shiitake are all dried commercially for the fine, concentrated flavour the process imparts. The liquid you use to reconstitute the mushrooms will absorb the flavour admirably.

To dry suitable mushrooms, place them over a source of dry and none-too-fierce heat such as a radiator or in an extremely low oven. Make sure they aren't touching each other or they may 'meld'. Many species can be threaded on fine string and suspended over an appropriate heat source. Cut larger specimens into sections or slices as appropriate.

Really keen foragers might like to invest in a fruit drier, available on the Continent for drying sliced apricots etc., but ideal for mushrooms. Tiered wire racks are stacked above a gentle source of heat.

Store the dried mushrooms in clear jars with tight-fitting lids. They make interesting displays in the kitchen.

The drying process also permits any 'foreign bodies' such as bugs and grit to detach themselves from the fruit bodies. When you reconstitute the mushrooms in warm water, look out for such debris. Use a fine strainer, or discard the last few drops of soaking water, which may be gritty.

Pickling

Pickling is particularly successful with firm-textured species like the Saffron Milk Cap. It probably makes sense to use this method for your best specimens.

200ml/7fl oz white wine vinegar
50ml/2fl oz balsamic vinegar
A little salt
1 tsp peppercorns
2 bay leaves
A little fresh thyme and dill
250–300g/around 10oz trimmed mushrooms

Simmer all the ingredients apart from the mushrooms for 15 minutes, then add the mushrooms and simmer on for a further 10 minutes.

Sterilize a jar by leaving in a warm oven for 30 minutes or by immersing for a couple of minutes in boiling water.

Pour in all the ingredients, seal and leave to pickle.

Freezing

Some, but not many, varieties of mushroom can be frozen directly after picking. Ceps, for instance, are commonly frozen while in their button stage. Caesar's Mushrooms – if you are lucky enough to find them – work perfectly too.

Other varieties freeze more successfully after they have been blanched, among them the White Button Mushroom, the Horn of Plenty and the Chanterelle.

Still other mushrooms benefit from being fried in butter – often the first step in a recipe – before being frozen. Winter Chanterelles are perfect for this treatment, as are Saffron Milk Caps and Hedgehog Mushrooms.

Stocks and Concentrates

Mushroom Stock

If you have been out collecting and find yourself with some specimens that aren't quite up to scratch – perhaps maggoty, or slightly nibbled at the edges – you can still use them to flavour up a stock nicely. Some mushrooms are better than others – Ceps, for instance, are fantastic. Making a stock is also a good way to capture the flavour of fungi which are difficult to preserve in other ways, such as Puffballs.

This stock will keep for up to 5 days in the fridge, or for up to 6 months in the freezer. It makes a great base for sauces and soups.

To make 1.75 litres/3 pints:
1kg/about 2lb mixed mushrooms
1 onion, roughly chopped
2 carrots, roughly chopped
2 sticks of celery, roughly chopped
2 cloves garlic, crushed
A sprig of parsley
A sprig of thyme
Salt and pepper, if desired

Place all the ingredients in a large pan and add enough water to cover, plus a little bit extra.

Bring to the boil and simmer for an hour and a half.

Strain through a fine sieve, taking care to leave any grit at the bottom of the pan.

Allow the stock to cool before putting it in the fridge or freezer.

Mushroom Extract

This is taking matters one step further than the stock recipe on page 155. The end result is so concentrated it is effectively a condiment.

The taste of such extracts can vary enormously according to the varieties of mushroom used in their creation. It would be impossible, for instance, to mistake Shiitake extract for extract of Cep. We would strongly recommend trying this process with both species to find out why.

To make 500ml/1 pint:

1kg/about 2lb mushrooms of your choice
1 litre/1¾ pints water
50ml/2fl oz soy sauce
100ml/4fl oz sherry
A small bunch of parsley
A small bunch of thyme

Place all ingredients in a large pan and bring to the boil. Simmer for an hour.

Pass the liquid through a muslin or very fine sieve, squeezing out every last drop.

Return the liquid to the pan, and continue to boil until reduced by a half to three quarters.

Store in the fridge in a sterilized jar (see recipe for Truffle Honey, page 109[?], for instructions on how to do this).

growing your own mushrooms

Duxelles

This is a great way of preserving your mushrooms, especially for the freezer. A 'duxelle' is simply a finely chopped vegetable, which, in this case, is cooked up with garlic, shallots and herbs.

This technique works well with many different kinds of mushroom. You may want to experiment with various mixes as well as making 'single-species' duxelles, for instance using Chanterelles.

50g/2oz unsalted butter
2 shallots, peeled and finely chopped
2 cloves garlic, chopped
500g/1lb of mixed mushrooms, finely chopped
Salt and pepper
50ml/2fl oz unsalted chicken or vegetable stock
A sprig of chopped parsley
A few more herbs, if you like, such as thyme, chives or dill

Melt the butter in a pan and fry the shallots and garlic gently until softened.

Add the mushrooms and a little salt and pepper. Fry until they have released their juices.

Add the stock and reduce until there is very little left at the bottom of the pan. Then stir in the herbs.

Either store in the fridge for a couple of days or freeze in small blocks in the freezer.

One solution to vagaries in the supply of mushrooms is to start growing your own. It is incredibly satisfying to harvest species which hitherto could not be 'harnessed'. Kits are now sold commercially for the home production of an ever-increasing number of species, including Parasol Mushrooms and Chicken of the Woods. In some cases results are virtually guaranteed, whereas in others (notably Morels) we are more in the territory of 'encouragement'. There is a growing trend towards 'Mycological Landscaping' – cultivating exotic mushrooms as beneficial companions to plants in gardens and lawns.

As producing mushroom spawn is a complex procedure involving specialist sterile culture skills, you will probably want to buy yours in. The two major formats are 'patches', where the mycelium arrives already mixed into its growing substrate, and 'plugs', which are inserted into holes drilled in logs of an appropriate species. Some species, such as Oyster Mushrooms and Shiitake, are available in both forms. The patches will grow indoors at normal room temperature.

It is also possible to buy 'Mushroom Stones' (see page 116), and saplings impregnated with Truffle mycelium.

Suppliers of mushroom kits include:

Fungi Perfecti
PO Box 7634
Olympia
WA 98507
USA
00 1 800 780 9162 or
00 1 360 426 9292
www.fungi.com

Mycologue
47 Spencer Rise
London NW5 1AR
020 7485 7063
www.mycologue.co.uk

Jac-by-the-Stowl
Penrhiw House
Llanddeusant
Llangadog
Carms SA19 9YW
Wales
01550 740 306
www.jac-by-the-stowl.co.uk

sources of further information

Bibliography

A Cook's Book of Mushrooms (Artisan, 1995) Jack Czarnecki

Fungi: Folklore, Fiction and Fact (Richmond Publishing Company, 1982) W.A.K. Findlay

The Mushroom Feast (Lyons and Burford, 1992 edition) Jane Grigson

A Gourmet's Book of Mushrooms and Truffles (HP Books, 1991) J. Hurst and L. Rutherford

The Ultimate Mushroom Book (Lorenz Books, 1995) Peter Jordan and Steven Wheeler

The Practical Mushroom Encyclopedia (Southwater, 2000) Peter Jordan and Stephen Wheeler

Mushrooms, Moulds and Miracles (John Day Company, 1967) Lucy Kavaller

The Fifth Kingdom (Second edition, Mycologue Publications, Canada, 1992) Bryce Kendrick

The Mushroom Book (Dorling Kindersley, 1996) Thomas Laessoe and Anna Del Conte

The Mushroom Identifier (Grange Books 1992) David Pegler and Brian Spooner

Mushrooms and other Fungi of Great Britain and Europe (Pan, 1981) Roger Phillips

In the Company of Mushrooms (Harvard University Press,1997) Elio Schaechter

Growing Gourmet and Medicinal Mushrooms (Third edition, Ten Speed Press, 2000) Paul Stamets

The Mushroom Cultivator (Agarikon Press, 1983) Paul Stamets and J.S. Chilton

A Morel Hunter's Companion (Thunder Bay Press, 1995) Nancy Smith Weber

Websites

Fungi Pefecti www.fungi.com (all sorts of mushroom-related stuff)

Mycosource www.mycosource.com (information on mushroom cultivation etc.)

British Mycological Society www.ulst.ac.uk/faculty/science/bms (information on local groups and societies, plus much more besides)

Suppliers

Mycologue
47 Spencer Rise
London NW5 1AR
Online mushroom shop. Contact Martin Lewy
Tel: 020 7485 7063
Fax: 020 7284 4058
Website: www.mycologue.co.uk

Jac-by-the-Stowl
Penrhiw House
Llanddeusant
Llangadog
Carmarthenshire SA19 9YW
Wales
A range of dried and fresh mushrooms plus growing kits and other mushroom-related paraphernalia
Tel: 01550 740 306
Website: www.jac-by-the-stowl.co.uk

Wild Harvest Ltd
31 London Stone Estate
Broughton Street
London SW8 3 QJ
Suppliers of exotic mushrooms to the restaurant trade. Also sell other wild foods.
Tel: 020 7498 5397

index

p13 'The Resurrection of Santa Claus' featuring *Amanita Muscaria*, by Jimmy Bursenos (2000), reproduced by permission of Jimmy Bursenos © Solstice Studios.
p15 'Blue Caterpillar' from *Alice in Wonderland* by Lewis Carroll. Illustrations by Sir John Tenniel, coloured by Diz Wallis, reproduced by permission of Macmillan Children's Books, London.
p16 'Caesar's Amanite Champignons', from *Les Champignons* by Dauphine (1900), reproduced by permission of Mary Evans Picture Library, London.
p21 'Organicus Caesareus', from *Les Champignons* by Cordier (1876), reproduced by permission of Mary Evans Picture Library, London.
p36 Netsuke Ivory Carving © Christie's Images Ltd.
p69 'The Feast' © Private Collection/Bridgeman Art Library
p101 'Truffle Hunters, Piedmont' by an unnamed artist, from a Leibig card, reproduced by permission of Mary Evans Picture Library, London.
p113 'Pig Finds Truffles' by an unnamed artist, taken from *Les Champignons Comestibles et Vénéneux* by Rocque, reproduced by permission of Mary Evans Picture Library, London.
p117 Mushroom Money Box © V&A Picture Library, V&A Museum, London.
p143 *Laetiporus* or Chicken of the Woods, reproduced by kind permission of the photographer Rod Talboys.